THE BIG QUESTIONS
IN SCIENCE

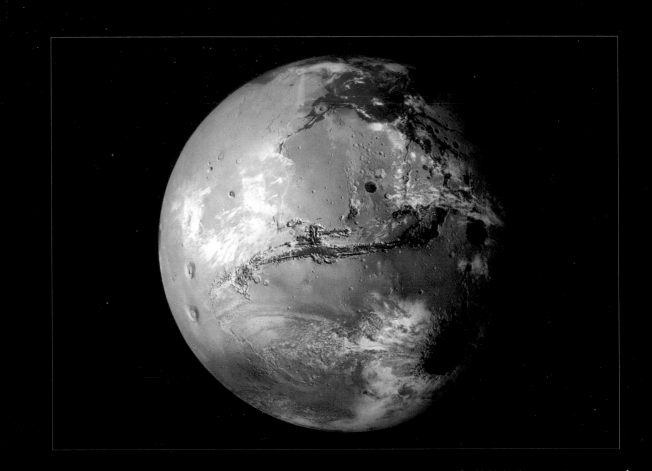

Could this be the answer to our population problem? A computer artwork shows what Mars would look like following a future terraforming. The artwork could also represent the planet in the past when its atmosphere was thicker and warmer and it had liquid water on its surface.

METRO BOOKS
New York

An Imprint of Sterling Publishing
1166 Avenue of the Americas
New York NY 10036

METRO BOOKS and the distinctive Metro Books logo are trademarks of Sterling Publishing Co., Inc.

Design © 2013 by Carlton Books Limited
Text © 2013 by Hayley Birch, Mun Keat Looi, Colin Stuart

ISBN 978-1-4351-6091-0

For information about custom editions, special sales, and premium and corporate purchases, please contact Sterling Special Sales at 800-805-5489 or specialsales@sterlingpublishing.com.

Manufactured in China

2 4 6 8 10 9 7 5 3 1

www.sterlingpublishing.com

THE BIG QUESTIONS
IN SCIENCE

THE QUEST TO SOLVE THE GREAT UNKNOWNS

HAYLEY BIRCH, MUN KEAT LOOI, COLIN STUART

METRO BOOKS
New York

HAYLEY BIRCH is a science writer and editor with a sideline in quirky science communication projects. She writes for popular science books and magazines, reports on environmental policy, produces podcasts, runs an academic comedy night and manages a science-themed solar-powered stage at a music festival in the Welsh hills. Based in Bristol, near Banksy's "Mild Mild West" mural, she inhabits a studio full of artists. She is an average endurance runner and a terrible saucier.

MUN KEAT LOOI is a science writer, editor, and all-round nerd. He's done stuff for the *Guardian,* Thomson Reuters, SciDev.Net and the Nuffield Council on Bioethics, among others, and is currently Online Editor at the Wellcome Trust, Europe's biggest biomedical research charity, running their online publications, blogs and social media. He lives in London, dreaming of Japan.
munkeatlooi.com

COLIN STUART is a space geek, astronomer and planetarium presenter. Also a freelance science writer, he has written for the *Guardian,* European Space Agency and BBC *Sky at Night* magazine. A fellow of the Royal Astronomical Society, he has talked about the wonders of the universe on TV and radio and presented from the top of the UK's biggest radio telescope. He lives with his wife in London and tweets as @skyponderer.

CONTENTS

INTRODUCTION – 6

1: WHAT IS THE UNIVERSE MADE OF? – 8

2: HOW DID LIFE BEGIN? – 18

3: ARE WE ALONE IN THE UNIVERSE? – 28

4: WHAT MAKES US HUMAN? – 38

5: WHAT IS CONSCIOUSNESS? – 48

6: WHY DO WE DREAM? – 58

7: WHY IS THERE STUFF? – 66

8: ARE THERE OTHER UNIVERSES? – 74

9: WHERE DO WE PUT ALL THE CARBON? – 82

10: HOW DO WE GET MORE ENERGY FROM THE SUN? – 90

11: WHAT'S SO WEIRD ABOUT PRIME NUMBERS? – 100

12: CAN COMPUTERS KEEP GETTING FASTER? – 108

13: WHEN CAN I HAVE A ROBOT BUTLER? – 116

14: HOW WILL WE BEAT BACTERIA? – 126

15: WILL WE EVER CURE CANCER? – 136

16: WHAT'S AT THE BOTTOM OF THE OCEAN? – 144

17: WHAT'S AT THE BOTTOM OF A BLACK HOLE? – 154

18: CAN WE LIVE FOREVER? – 164

19: HOW DO WE SOLVE THE POPULATION PROBLEM? – 172

20: IS TIME TRAVEL POSSIBLE? – 182

INDEX – 190   CREDITS – 192

INTRODUCTION

"Prudens quaestion dimidium scientiae"
(To ask the proper question is half of knowing)

A saying attributed by the philosopher Roger Bacon to the
the Greek philosopher Aristotle

Humans have always asked questions. Our desire to learn about our world and our humanity is a defining aspect of our nature. Sometimes we ask because we need to know the answers, but often just because we're interested enough to want to know. Indeed, questions form the very basis of science and the scientific method: pose a question, come up with a theory, carry out experiments to find out whether you're right.

In pursuit of answers, we stray down unfamiliar paths, pushing the boundaries of what we can achieve and often leading to unexpected discoveries, epiphanies and new technologies. In 1869, the Irish physicist John Tyndall wondered why the sky was blue. In working out the answer – related to the way air particles refract light – he inadvertently laid the foundations for lasers and fibre optics. He also discovered that bacteria can't grow in pure air, banishing the myth that germs appear spontaneously, and that the airways of the lungs remove particles from the air that we breathe. Curiosity-driven "blue skies" research not only leads us to answers; the journey can be just as enlightening.

Questions come in all shapes and sizes. The biggest ask what we are (What makes us human?), why we are (Why is there stuff?), what our world is (What is the universe made of?) and how we will survive it (Can I live forever?). There are also questions that speak to our sense of adventure, the spirit of endeavour that pushes us to the furthest reaches of this planet (What's at the bottom of the ocean?), this universe (Are there other universes?) and this existence (Is time travel possible?). Few of these questions are easy to answer; many are very, very hard. Some are unanswerable because there is no way of knowing if our answers are right (How did life begin?), some because there isn't one right answer

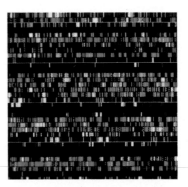

A computer display of a human DNA sequence from the Human Genome Project. The coloured bands represent a sequence of bases in our genetic code. Studying this code is giving us a greater understanding of genetic diseases, heredity, and who – and what – we are.

(What is consciousness?), and some because we don't yet have a way to find the answer (What's at the bottom of a black hole?). Of course, politics and economics can get in the way as well (How do we solve the population problem?).

The answers to these big questions have eluded us for centuries, millennia even. And the quest to answer them continues, taken up by a cast of hundreds, sometimes thousands, from humble laboratory scientists to extravagant millionaires – some you will have heard of, many you will not. Biologists, chemists, physicists, mathematicians, computer scientists, philosophers, writers, explorers and engineers – all play a part, fuelled by their natural curiosity, united over great distances and history by a desire to answer questions.

This book provides just a small sample of the many questions that science is still working to answer, and just a fraction of the many fascinating solutions and theories that scientists have come up with so far. The wonderful – and maddening – thing about questions is that there is no end to them; answering one inevitably poses others. That's what keeps scientists searching for answers and humans curious. As Albert Einstein once said:

"Learn from yesterday, live for today, hope for tomorrow. The important thing is to not stop questioning."

Scientists are searching for new ways to power our world. Even the 72,000 solar panels at the Nellis Air Force Base in Nevada provide only a quarter of its energy needs. So the question is: how do we get more energy from the Sun?

1

WHAT IS THE UNIVERSE MADE OF?

Look around you. Everything you can touch and see is made of atoms. This book, the chair you are sitting on, the socks you put on this morning. The sky, too, is filled with atoms – those in the atmosphere scatter the blue parts of sunlight most, giving the sky its distinctive colour. The Sun itself, along with its family of planets, moons, comets and asteroids, is made up of these tiny building blocks, as are the myriad of stars that are spread across the vastness of space. So you might be forgiven for thinking that the question of what the universe is made of has already been answered. But in recent years astronomers have come to a startling realization: atoms – the stuff of people, planets and stars – make up a mere 5 per cent of the universe. So what about the remaining 95 per cent? The hunt for this "missing matter" is one of the hottest topics in modern physics.

Since the seventeenth century, scientists pondering the size and shape of the universe have faced a problem. Isaac Newton had unveiled the laws of gravity in which anything with mass is attracted to everything else with mass. We stay on the ground because our mass is attracted to the mass of the Earth. But if everything in the universe is mutually attracted to everything else, then eventually the universe should collapse in on itself as its contents are drawn closer together. That clearly wasn't happening. The problem remained unresolved by the time Albert Einstein published his general theory of relativity in 1915, and by 1917 he had been forced to alter his equations to include a buffer against gravity. Assuming everything in the universe stays where it is, he needed this buffer to keep things balanced and preserve the static universe that he believed to exist. He called this buffer in his equations the cosmological constant. However within a few decades measurements revealed the universe isn't static at all.

In 1929 US astronomer Edwin Hubble made one of the most groundbreaking discoveries in the history of astronomy: the universe is expanding. Hubble was surveying galaxies – aggregations of stars, gas and dust that form bright islands in the blackness of space. He was able to measure both the distance to a galaxy and the speed at which it was moving. As a galaxy moves away from us, its light gets stretched out, shifting it towards the red end of the light spectrum. It is from this apparent change in colour that the effect get its name: redshift. The more the light from a galaxy is redshifted, the faster it is moving away. What Hubble found shook the world of physics. The further away a galaxy the faster it appeared to be moving away from us – exactly what would be expected in an expanding universe. Realizing his mistake, Einstein called the addition of the cosmological constant his "greatest blunder".

The expanding universe threw up some immediate questions. If the universe is getting bigger over time, then it must have been smaller yesterday, and smaller still the day before. The galaxies and the matter they contain must all have been closer together in the past. At some point long ago, all the matter must have been concentrated in one place – the

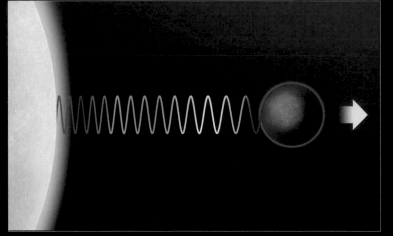

universe must have had an origin. Astronomers call this the Big Bang, and by using mathematics to rewind the expansion they can date this beginning to around 13.8 billion years ago. For a long time, they believed that the rate of this expansion was slowing down as the gravitational attraction between everything in the universe gradually applied the brakes. However, in 1998 astronomers' ideas about the universe were once more turned upside down.

Since the late 1980s, two teams of astronomers had been studying variations in the expansion rate of the universe over its history. They had been looking at Type 1a supernovas, some of the brightest phenomena in the universe. This type of supernova occurs when two stars are in orbit around each other and one of the stars has already died. The dense core of the dead star (known as a white dwarf) starts to rip gas from its neighbour, slowly gaining more and more mass. However, once the white dwarf reaches a certain mass – about 1.4 times the mass of our Sun – it becomes unstable and explodes. As it always explodes with the same amount of fuel (1.4 solar masses), the supernova always has the same intrinsic brightness. But because light fades over distance, it will appear dimmer when viewed from Earth, and the amount of dimming allows astronomers to work out the distance to the exploding star. As Type 1a supernovas are so bright, they can be used to measure distances across 75 per cent of the visible universe. Furthermore, because light takes time to travel through space, the further away the supernova is from us, the further back in time astronomers are peering. So by looking at the redshifts of galaxies containing these supernovas, astronomers can look back into the past and measure the rate of the universe's expansion at the time when the supernova exploded. By 1998, both teams had arrived at the same conclusion: the expansion of the universe is now speeding up, not slowing down. From their measurements, it looks like the expansion of the universe stopped slowing between five and seven billion years ago and then started accelerating. For this discovery, US astronomers Saul Perlmutter, Brian Schmidt and Adam Reiss were jointly awarded the 2011 Nobel Prize for Physics.

DARK ENERGY

Because the expansion of the universe is speeding up, something must be acting as a kind of "anti-gravity", pushing galaxies ever further apart. The true nature of this mysterious entity is still unknown, and for now it is simply called dark energy. Even though astronomers know little about the nature of dark energy, by using measurements of how fast the galaxies are moving apart, they can estimate how much dark energy there must currently be. The best estimates have dark energy making up a staggering 68 per cent of all the stuff in the universe. So what could dark energy be? What could counteract gravity in this way? One, perhaps surprising, idea is space itself.

Physicists already know that there is no such thing as empty space; even in a vacuum there are energy fluctuations. And since space pervades the whole universe, so should this "vacuum energy". However, it gets stranger, because the laws of physics say that if a vacuum always has a certain amount of energy, then this must be balanced by a repulsive gravitational effect. What's more, the overall amount of this repulsive vacuum energy, per little piece of space, remains constant no matter how much space expands. Perhaps dark energy, the accelerating expansion of the universe, is due to the fact that the repulsive force of vacuum energy

- - →

Type 1a supernova 1994D (bottom left) briefly outshines its fellow stars in the NGC 4526 galaxy. Measurements of these

be that the expansion of the universe started accelerating when the strength of the galaxies' gravitational attraction dipped below the constant repulsive force of vacuum energy. It is like a tug of war where the members of Team A always remain in the same positions while their opponents on Team B start closer together but are forever spreading out. Initially Team B has the advantage but eventually Team A will win the battle.

Remarkably, this constant vacuum energy density, with its repulsive properties, is exactly equivalent to the cosmological constant that Einstein added to his equations of general relativity back in 1917. Perhaps Einstein didn't blunder after all and dark energy is already there in the equations of general relativity, albeit by mistake. If dark energy really is just vacuum energy then it, too, should be constant and unchanging. Fortunately, it is possible to test this. Astronomers can measure what is known as the "equation of state" of dark energy – the ratio of the pressure of dark energy to the amount of stuff in a fixed volume of space. If the value of this equation of state is exactly -1, then dark energy is Einstein's cosmological constant. However, if the equation of state is variable – if astronomers can find evidence of it ever differing from -1 – then dark

Artist's impression of the Big Bang. Until dark energy was discovered, it was thought that gravity was slowing the expansion of the universe caused by the force of the Big Bang. Astronomers now know the expansion is getting faster.

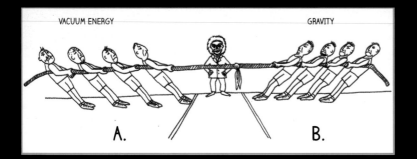

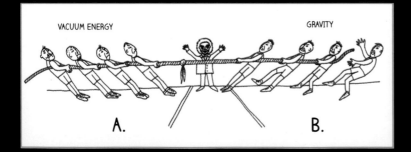

Einstein said he "blundered" by introducing a cosmological constant into his equations in 1917. But today many believe it is Einstein's constant that is providing the driving force behind the universe's accelerating expansion.

- - →

energy isn't constant and it is less likely to be explained by Einstein's "blunder". Current experiments do, indeed, seem to indicate that the equation of state has a value of -1 and so, at least for now, Einstein appears to have been right to introduce a cosmological constant into his equations of general relativity.

However, there is a problem. The total amount of vacuum energy researchers expect to exist is vastly greater than the energy required to drive the current expansion of the universe. In fact, it is greater by a factor of 10^{120} (1 followed by 120 zeroes)! Nevertheless, it is the best answer that we currently have. So the hunt for the true nature of the missing 68 per cent of our universe will continue, with new space missions and telescopes to help probe the mystery further.

DARK MATTER

The present-day universe is made up of 5 per cent atoms and 68 per cent dark energy, but that still leaves 27 per cent – more than a quarter – unaccounted for. Researchers believe this remainder is another shadowy entity, first discovered by Swiss physicist Fritz Zwicky in 1933. Zwicky, one of science's most eccentric characters, was using the telescopes of Caltech University in the USA to look at the Coma Cluster – a collection of around 1,000 galaxies situated some 300 million light-years (about 2,850 trillion kilometres, or 1,740 trillion miles) from Earth. He was able to measure the redshift of the galaxies in the cluster and realized that they were moving much faster than they should be. The gravity of all the material he could see in the cluster wasn't enough to account for the galaxies' speed, so he suggested that there must be something else in the cluster he couldn't see but whose gravitational influence was providing the galaxies with the extra speed. He called this murky material Dunkelmaterie, the German for "dark matter".

A few years later, in 1937, US astronomer Sinclair Smith found a similar phenomenon in the Virgo Cluster. However, his work is less remembered because he died of cancer a year later, aged just 39. Around the same time, Dutch astronomer Jan Oort was investigating the movement of stars within our own galaxy, the Milky Way. He observed

Mosaic of the Coma cluster, combining visible and infra-red images from the Sloan Digital Sky Survey and Spitzer Space Telescope. One of the first signs of dark matter came in 1933, when Fritz Zwicky realized these galaxies were moving too fast.

that some stars were moving fast enough that they should escape from the Milky Way entirely. However, they clearly weren't and so something must be holding on to them.

Despite the findings of Zwicky, Smith and Oort, the subject of the "missing mass" stayed on the fringes of astronomy until the 1970s. Then, US astronomer Vera Rubin noticed a problem with the Andromeda Galaxy, the nearest major galaxy to our own Milky Way. The stars near the edge of the galaxy were rotating much too fast. Most of the visible material in flat, spiral galaxies like Andromeda and the Milky Way seems to be concentrated in the central bulge, with the rest of the stars orbiting around this region. What Rubin expected to find was that the further out in the galaxy she looked, the slower the stars would be moving around the centre. This is just the same as we see with the planets in our own solar system: the inner planets Mercury, Venus, Earth and Mars orbit the Sun quicker than Jupiter, Saturn, Uranus and Neptune, which lie further out. What she actually found was that the rotation speed of the stars remained almost constant across the whole galaxy. Although it appeared

An ultraviolet view of the Andromeda galaxy through the eyes of the GALEX space telescope. In the 1970s, US astronomer Vera Rubin noticed stars towards the edge of the galaxy were moving faster than expected. This can be explained if unseen dark matter is spread throughout the galaxy.

that most of the mass was concentrated in the centre, like it is with the Sun in our solar system, the stars in Andromeda didn't behave like the planets. By 1980, Rubin had published details of the same phenomenon in over 100 galaxies.

The findings of all four astronomers can be explained if there is some extra material that doesn't interact with light, and so isn't visible, but that still exerts a gravitational influence. In the case of Rubin's rotating galaxies, if this unseen material is spread throughout the galaxy, the overall mass would not be concentrated in the centre and things wouldn't behave the same as in our solar system. Similarly, the pull of the same unseen material would keep Oort's stars from flying off into intergalactic space and speed up the galaxies in the clusters observed by Zwicky and Smith. This mysterious material continues to be known as dark matter because it seems to behave like normal matter in every way, apart from the fact that it doesn't emit, reflect or absorb light.

Scientists have continued to find evidence of dark matter throughout the universe, using techniques such as gravitational lensing. Lensing can happen when a big cluster of galaxies blocks our view of a galaxy behind it. So massive is the cluster that its gravity acts as a lens, bending some of the light from the background galaxy around the cluster. The light from the galaxy then appears as arcs around the edge of the cluster. When all the visible mass of the stars, dust and gas in the cluster is added up, it isn't enough to account for the amount the light is bent. The favoured explanation is that the cluster contains

Hubble Space Telescope image showing galaxy cluster 0024+1654 (yellow) acting as a gravitational lens. Light from a more distant galaxy is bent by the cluster's mass, appearing as arcs of light (blue). Visible matter often can't account for the total amount of lensing and so dark matter is needed to help explain it.

← – –

INSIDE EXPERT:
WHAT ARE WIMPS?

"As the case for dark matter strengthened in the 1980s, panicked astronomers turned to particle physicists, asking what the invisible stuff might be made out of. But the particle physicists had their own concerns: they needed to explain odd coincidences in the mathematics of normal, everyday particles (the 'standard model').

One new theory, supersymmetry, seemed able to tackle both problems. The theory supposes that each normal particle type has a massive 'super-partner' – for instance, quarks have super-friends called squarks. For particle physicists, these super-partners iron out peculiarities in the standard model of normal matter. For astronomers, the most stable of the super-particles behave delightfully like dark matter should.

First, the particles don't reflect, absorb or emit light. That makes them invisible, which is a good start. Second, the particles should very occasionally bump into atoms of normal matter (so they're Weakly Interacting Massive Particles' – WIMPs). That makes the theory testable, which is always helpful. Third, you can calculate how many WIMPs should have been manufactured in the Big Bang: neatly, it's just the right number to account for dark matter.

The problem with having a testable theory, though, is that people will insist on testing it. And, as of early 2013, there's no sign of supersymmetry at the Large Hadron Collider. That's not yet a death knell for the theory, but if nothing turns up in the next few years, we may have to revise our best guess at the culprit behind dark matter."

Dr Andrew Pontzen,
Theoretical Cosmologist, University of Oxford, UK

enormous clumps of dark matter that not only bind the cluster together but whose gravity contributes the extra needed to account for the amount of light bending.

However, despite the increasing amount of evidence for its presence throughout the universe, dark matter has never been detected directly; its presence has only been inferred from effects on other, visible material. Many teams of scientists around the world are working on direct-detection experiments in an attempt to find concrete evidence about dark matter's true nature. The leading idea is that dark matter takes the form of Weakly Interacting Massive Particles (WIMPs) (*see* INSIDE EXPERT: WHAT ARE WIMPS?). Several detectors around the world are looking for evidence of WIMPs passing through the Earth while others are looking for the effects of interactions between WIMPS out in space.

Whatever dark matter and its shadowy cousin dark energy turn out to be, they make up a huge proportion of our cosmos. No longer do we think we live in a universe made of the same stuff we are. And yet it is from our little blue planet, a world made of these rare atoms, that we are able to look out into space, ask the big questions about our universe and attempt to unveil the true nature of reality.

HOW DID LIFE BEGIN?

Almost anything that was once alive can be made into a decent pot of soup – cauliflower, seaweed, pig's trotters, even fruit bats (on certain islands in the west Pacific). The American chemist Stanley Miller appears to have taken this idea and turned it on its head. He believed that soup was the origin of everything that ever lived. In one of the iconic science experiments of the twentieth century, Miller cooked up his famous "primordial soup", which was nothing less than a recipe for life itself. He took the chemical ingredients he supposed were available on an earlier, warmer Earth and transformed them into a hot broth loaded with biological molecules. Some of these same molecules are present in every living thing today, implying that the evolutionary path leading all the way to us began with the specks of life that floated in the primordial soup some four billion years ago. When Miller published his results in 1953, he was just 22 years old and had the origin of life figured out. Well, not quite. The problem, of course, is that because no one was there to witness the event itself, we are left to wonder what really happened during that soupy beginning.

Miller was not the first to accept the notion of life emerging in a chemical soup. In 1871, an ageing naturalist by the name of Charles Darwin wrote to his friend, botanist Joseph Hooker, about a certain "warm little pond". In it, Darwin imagined the effects of light, heat and electricity turning the few basic chemicals available on our primitive planet into proteins – the biological molecules responsible for digesting our food and making our muscles move, as well as millions of other jobs. Darwin later dismissed these ponderings, saying it was "mere rubbish" to think of the origin of life at a time when there was so little scientific grounding.

It was a Russian biochemist, Aleksandr Ivanovich Oparin, who influenced Miller. In the 1920s, Oparin brewed up the idea of the prebiotic soup, which he published in 1924 in his book *The Origin of Life*. Based on what was then known about the atmosphere of Jupiter, and assuming our own planet might once have had a similar atmosphere, the scene he envisaged for early Earth was a suffocating oxygen-less mixture of methane, ammonia and hydrogen gases, amid a mist of water vapour. He believed that these simple components reacted to form complex carbon-based molecules that coalesced in the oceans and later became abundant in cells. And so the soup theory thickened, creating a stir among Darwin's followers, and eventually seeping into the consciousness of Miller as an eager young teaching assistant at the University of Chicago. Here, he persuaded his PhD supervisor – who just happened to be the Nobel Prize-winning chemist Harold Urey – to let him carry out the experiment that would test Oparin's theory. The set-up consisted of a couple of glass flasks and Oparin's list of primordial ingredients. Miller connected the two flasks and filled the bottom one with a boiling "ocean" of water that evaporated into the gas mixture in the top flask. The drama of the classic experiment was heightened by the addition of a Tesla coil discharging electric sparks to simulate lightning in the atmosphere of early Earth.

What Miller found floating in his soup a few days later proved that it was possible for the complexity of biology to arise from simple chemistry. A series of smudgy stains on a piece of chromatography paper signified the presence of amino acids – the subunits that link together to make proteins in the bodies of all living things, from amoebas to zebras. But Miller was far from finding life's beginning. For a start, there are some basic problems with the view of the world – originally Oparin's – that he set out to test. Most scientists now working in the field agree that carbon dioxide and nitrogen were important components of Earth's early, prebiotic atmosphere, yet neither featured on Miller's ingredient list. Perhaps most intriguingly though, the primordial soup course only serves to whet the appetite for the main course ahead. Chemicals alone, no matter how complex and rich in carbon, do not constitute life. What came after soup?

Taken on 16 May 1953, this photograph shows a young Stanley Miller in his lab at the University of Chicago the day after his classic paper on the origin of life was published.

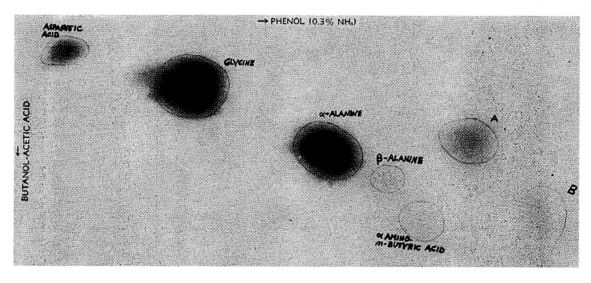

↑ Soup spots. Using a piece of chromatography paper, Miller separated out some of the components of his primordial soup experiment. Amino acids are circled, showing that complex biological molecules had formed.

Cells, by all accounts, came after soup. The fact that life exists as cells at all, instead of as one giant, gloopy mass of molecules, tells us something about a fundamental requirement for life: it needs containment. In single cells like bacteria, the fatty layer surrounding each cell creates the compartment that life needs to separate its chemical goings-on from those of other bacteria. In life forms made of millions or even trillions of cells, like people, compartmentalization also allows teams of cells to specialize in doing different jobs. We know that cells appeared at some point after soup, but in the meantime another important requirement for life had to be fulfilled: replication. When our cells divide, they make copies of the information they carry in their genes and pass them on to their daughter cells, so that the daughter cells turn out like their parent cells. In the beginning, molecules must also have found a way to pass on information. They did not have the complex machinery of the entire cell at their disposal to help them, so some of the molecules floating in the primordial soup must have copied themselves – they must have been self-replicators.

Scientists were seeking a molecule that was simple enough to have formed in the primordial soup yet had the ability to make perfect copies of itself unassisted. Could proteins, the workhorses of cells, have accomplished such a feat? As Miller was boiling up his broth in Chicago, two biologists at the Cavendish Laboratory in Cambridge, England, were pinning down another key piece of the puzzle. A few weeks before Miller published the results of his prebiotic soup experiment in 1953, James Watson and Francis Crick unveiled their double helix structure for deoxyribonucleic acid (DNA) – the molecular blueprint for life. The era of genetic research that followed this breakthrough discovery eventually led to ribonucleic acid (RNA) – DNA's smaller, single-stranded cousin – becoming the prime candidate as the original self-replicator, the molecular origin of life. Today, nucleic acids – DNA and RNA, the molecules that carry the genetic code to make proteins – are vying with proteins themselves over the question of which came first. Was it nucleic acids, the information-bearers, or proteins, the molecules that do, well, just about everything?

CHICKEN OR EGG?

The "which came first?" problem of nucleic acids versus proteins is made all the more difficult to solve by the fact that the fates of nucleic acids and proteins are irrevocably entwined. RNA, like DNA, is a long and complex molecule that carries a code in the form of the sequence of its subunits (called bases). RNA's job is to ferry copies of the genetic information encoded in DNA to protein-making factories in cells. All of the 100,000 or so different proteins going about their business in our bodies today would cease to exist without the instructions encoded in DNA and borne by RNA. But just as there can be no proteins without nucleic acids, there can be no nucleic acids without proteins. One of the many roles that proteins must perform is making new DNA and RNA. The problem is akin to the chicken and the egg riddle. So what has convinced some scientists that RNA might have beaten proteins to it as the original self-replicator? It is certainly difficult to imagine how a molecule as complicated as RNA could have emerged first from the prebiotic soup, particularly when considering the ease with which Miller appears to have created the forebears of proteins in a flask full of gas and steam.

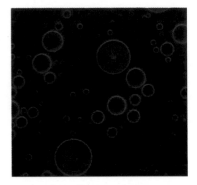

Lab-made "protocells" represent simple vessels for life – like early cells – but could also be used to create brand new forms of life.

The soup-to-cells story picks up with Walter Gilbert, who studied chemistry and physics at Harvard University but ended up as a molecular biologist – a series of events that is perhaps less surprising in light of the influence that James Watson had on his early career. First meeting during Gilbert's doctoral studies at Cambridge University, Watson and Gilbert were both back home in the US and at Harvard by the late 1950s. There, Watson turned his attention from DNA to RNA and Gilbert initially taught theoretical physics. But Gilbert became so excited by the experimental work Watson was doing that he decided to join him in the Harvard biology laboratory and spent the the summer of 1960 studying nucleic acids. He continued in the field of molecular biology, being awarded the 1980 Nobel Prize in chemistry (shared with Frederick Sanger and Paul Berg) for work on nucleic acids, and in 1986 christened the "RNA world" theory, in a paper published in the scientific journal *Nature*.

According to the RNA world theory, RNA, not protein, was the precursor of life. It was the first molecule to self-replicate, producing more RNA, and therefore the molecule that got evolution started.

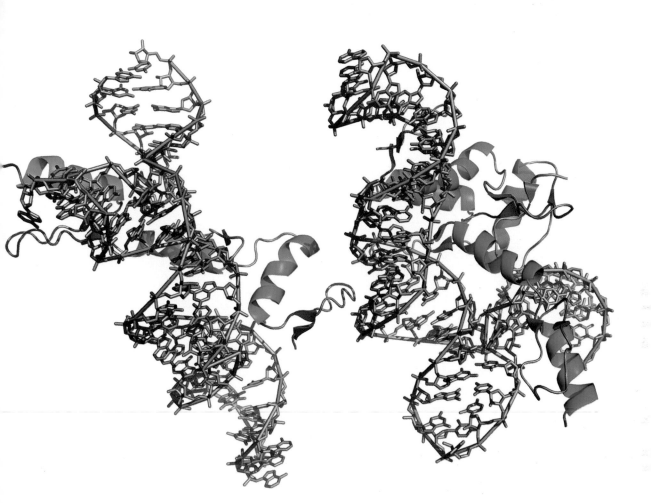

Ribonucleic acid (RNA) (pink) is a candidate for the molecule that started life on Earth. In cells, it carries copies of the DNA code to protein-making machines, where it interacts with "switch" molecules (blue) that determine whether the code is used as a pattern to create new proteins.

However, this theory has not been as warmly embraced by other scientists studying the origin of life as Gilbert might have hoped. It has gained the support of more scientists than other, competing theories but only grudgingly, having been described as the best of a bad bunch. Writing in 2012, New Zealand biochemist Harold Bernhardt referred to the RNA world hypothesis as "the worst theory of the early evolution of life (except for all the others)". Given that it is based on something that happened about four billion years ago, Gilbert's theory cannot be tested directly. Not only that, but scientists have not been able to find RNA that can self-replicate, or even a molecule like it. They have examined trillions of different RNA sequences looking for self-replicators, and although they have found molecules that can make abridged copies of themselves, they have not tracked down one that can make a full-length version of itself. But if RNA as the precursor of life is lacking experimental support, the principal alternative contenders, proteins, are hardly doing any better.

A recent tactic used by scientists has been to compare the evolutionary histories of RNA and protein to try to work out which is older. Just as it is possible to say that crocodiles evolved before cats by looking at information contained in their genes, it might also be possible to say that RNA is older than protein – or vice versa – by looking for information hidden in their

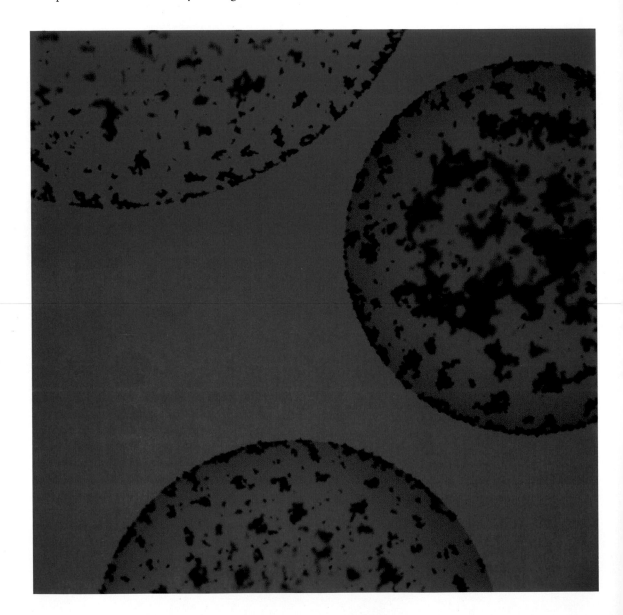

INSIDE EXPERT: WHAT WERE THE FIRST CELLS LIKE?

"A living cell is a system of molecules inside a membrane compartment, all of which are working together to make the life process occur. We don't know which particular set of components got together to make the first cell – the first system of molecules – but it seems likely the first cell membranes were already there. Like soap bubbles, these membranes would have been made from molecules called lipids, and we know from experiments that lipids can self-assemble into membranes in conditions similar to those on the early Earth four billion years ago. Then lots of other molecules got caught inside these microscopic bubbles or lipid vesicles. They were like microscopic test tubes, trying to figure out how to become alive by doing trillions and trillions of experiments. And I think that's how life began

– by the rare vesicle, out of these countless numbers that were being produced and broken down all the time, just happening to have a system of molecules that could capture energy and nutrients and begin to grow. The simplest form of cellular life now is a bacterium called *Mycoplasma*, but even this cell has hundreds of genes and makes thousands of different proteins. We're not at the point yet where we can say with any certainty what the first form of life was like except that it had to be much, much simpler even than the simplest forms of life today."

**Professor David Deamer,
biomolecular engineer,
University of California, Santa Cruz**

These oil droplets, containing simple chemicals, may resemble the first cells. The black compounds within are hydrogen cyanide polymers, which can be used to power protocell movement through chemical reactions.

molecular structures. The trouble with this approach is that it is more difficult to trace the ancestors of proteins than those of RNA, and this is because, in modern cells, RNA carries the code that makes the proteins. To appreciate why this is a problem, let us pretend for a moment that proteins, not nucleic acids, were the original self-replicators. It is four billion years ago and there are the proteins, duplicating themselves under the primordial Sun and making more and more proteins. Then along comes RNA to reinvent the whole system. Soon, proteins are trapped inside cells, being generated according to a set of instructions borne by the RNA. Whatever happened before the advent of the new RNA system is untraceable, because all the proteins in our cells today are based on the new system.

Even if we put aside the problem of whether RNA or proteins were the precursor molecules and agree that eventually they started to work together (which they did), there are still many other questions to resolve about the origin of life. For instance, where did the first cells come from and when did they first appear? What were they originally made from? (*see* INSIDE EXPERT: WHAT WERE THE FIRST CELLS LIKE?) These questions are leading us to create in the laboratory simple biological systems that mimic early cells – protocells. It is now also possible to synthesize DNA

and wrap it in a cell-like skin. In the near future, scientists may well achieve the ultimate goal of making a protocell that can self-replicate. Then they may claim that they have created life in the laboratory, but they will never be able to claim that they have uncovered the origin of life. That, it seems, is a secret only a time traveller could discover (*see* QUESTION 20: IS TIME TRAVEL POSSIBLE?).

Scientists are still arguing over the conditions that provided the cradle for life. Some believe Miller's hot primordial soup was actually cold – a kind of gazpacho – others that it was acidic, like vinaigrette, or oily, like mayonnaise. Meanwhile, there is still no general agreement about where, never mind how, life began. While some think life was kick-started by meteorites hitting the oceans, others argue that meteorites carried life to Earth from another planet, neatly dispensing with the problem of how life got started here and transferring it to some unknown alien race, perhaps living in a galaxy far away, pondering its own origins. A more likely story, though, is that life originated near volcanoes. It is almost certainly true that Earth was a lot hotter a few billion years ago – perhaps by many degrees – and its surface pocked by volcanic eruptions. For this reason, the primordial soup bowl is often thought of as the pool surrounding a hot spring, such as might occur near a volcano today. First life could even be linked to volcanic activity under the sea floor, at hydrothermal vents known as "black smokers", where

The average global temperature on Earth four billion years ago may have been as high as 70 degrees Celsius (158 degrees Fahrenheit). A popular theory is that life emerged in the springs and vapours surrounding volcanoes. Today, heat-loving (thermophilic) bacteria still thrive in thermal springs.

water heated by molten rock pours forth and evolutionarily ancient animals abound (*see* QUESTION 16: WHAT'S AT THE BOTTOM OF THE OCEAN?).

Given the wealth of new ideas about our origins, Miller's soup theory could be seen as obsolete. Perhaps it is time to dispense with the soup theory altogether and face the fact that we will never know for sure how life began. But Miller still had a trick up his sleeve. After he died in 2007, Miller's former student, geochemist Jeffrey Bada, found himself going through some boxes that Miller had left in his laboratory. They were stuffed full of vials – apparently Miller was something of a hoarder, keeping samples from each experiment, labelling them and jotting them down in his laboratory notebook. Bada was amazed to discover vials from Miller's experiments in 1953 and 1954, just after his primordial soup paper was published, and also from unreported studies in 1958. With the benefit of new equipment and new understanding, could Miller's original samples tell Bada anything he did not already know about the origin of life? It may have been asking a lot to expect 50-year-old samples to bring any new information to bear, but re-analyzing the contents of the vials did give Bada something to think about. The samples were from modified soup flask set-ups that mimicked volcanic mists – in one case using a nozzle to propel jets of hot steam into the flask – and appeared to contain a few more complex chemicals than Miller's laboratory was capable of identifying at the time. These forgotten experiments suggested that volcanic activity was crucial in generating biological complexity from chemical beginnings. So although Miller is not the only scientist to have considered volcanoes as the cradle of life, his work is still able to influence modern thinking on our origins.

The deoxyribonucleic acid (DNA) code is copied so that information can be passed from cell to cell. But DNA is too complicated a molecule to have seeded all of evolution, leaving us wondering if its simpler cousin RNA could have done the job. – – →

ARE WE ALONE IN THE UNIVERSE?

A robot sits on top of a vast ice plain that stretches to the horizon in every direction. Receiving the go-ahead, the machine begins to melt its way down. After travelling 20 kilometres (about 12 miles) through solid ice, the robot's job is done and it spawns a second machine capable of swimming around in the salty water beneath. With a light to illuminate the dark depths, this machine begins exploring this subterranean ocean. Its quarry are the tiny microbes that might reside here.

These hypothetical robots are not exploring the polar regions of Earth, but the subsurface oceans of Europa, a moon of the planet Jupiter. With vast oceans of liquid water lurking beneath enormous ice sheets, Europa is a prime location to look for life elsewhere in our Solar System. Future missions like this could revolutionize our understanding of life and begin to answer the question of whether we are alone in the universe.

But 750 million kilometres (466 million miles) from the Sun, this water should freeze in the piercing cold of the Solar System's outer reaches. Yet astronomers believe that there is more liquid water on Europa than in all the lakes, rivers, seas and oceans of the Earth combined. This water is sloshing around, shifting and cracking the ice sheets so that, from above, the surface of Europa resembles the top of a cracked crème brûlée.

The culprit responsible for the presence of water on this distant world is the immense gravity of Jupiter, the largest planet in the Solar System. The pull of Jupiter's gravity flexes Europa as it orbits, making it successively larger and smaller. In a similar way to repeatedly bending a piece of metal, the continuous flexing heats up the moon's core and provides enough energy to keep its water from freezing. And where there is water, there might be life. This warming by Jupiter – known as tidal heating – could even drive hydrothermal vents similar to those thought by some scientists to have kick-started life on our own planet (*see* QUESTION 2: HOW DID LIFE BEGIN?). The European Space Agency is planning to launch a probe, the Jupiter Icy Moons Explorer (JUICE), to further explore Jupiter's moons, including Europa, in 2022. If the mission goes ahead, it might help to establish whether the moons could, in fact, support life.

This isn't the first time astronomers have got excited about the link between water and alien life. In 1877, Italian astronomer Giovanni Schiaparelli announced his discovery of "canali" (channels) on Mars. He drew maps depicting a weave of intersecting straight lines spanning the planet not too dissimilar to the criss-crossing lines of Europa. But when his work was translated into English, the word "canali" was mistranslated as "canals". A seemingly innocent mistake, it had a profound effect — channels can be carved by nature whereas canals are constructed by intelligent creatures. Such was the fervour surrounding this discovery that there was an explosion in science fiction describing an inhabited Mars, with H.G. Wells' 1898 novel *The War of the Worlds* being the most famous example.

Criss-crossing cracks on the surface of Europa, a moon of Jupiter, caused by an ocean of salty water shifting beneath vast ice sheets.

The intense scientific speculation about the nature and existence of the Martian canals, and their potential intelligent creators, continued well into the twentieth century. The argument was finally settled in 1965, when the Mariner 4 spacecraft performed the first flyby of the planet and failed to find any evidence of the channels. Successors to Mariner 4 have orbited around, landed on and driven across Mars and have revealed it to be a dry, dusty and desolate place without any obvious signs of life. However, those successors, including the *Spirit*, *Opportunity* and *Curiosity* rovers, have also revealed that Mars was once a warm and wet planet with water flowing on its surface. Water can still be found on Mars today, but it is locked up in ice and can't flow freely for any length of time. It's possible that liquid water remains deep under the rusty crust of Mars and any life that once thrived in the planet's previously temperate conditions could still be clinging on in these underground aquifers. Ideas for future Mars exploration include a drilling mission to excavate this potential life-supporting region.

A late nineteenth-century drawing of Mars by Italian astronomer Giovanni Schiaparelli showing the apparent "channels" which, when mistranslated as "canals", sparked decades of speculation about intelligent life on Mars.

Despite the tantalizing possibilities of life on Mars or Europa, our own planet remains far and away the most desirable real estate in the Solar System when it comes to having the amenities required for life. A major reason for the abundance of life here is thought to be our snug temperature. Earth is said to be in the "Goldilocks Zone" – the small region around a star where, like the porridge in the fairy tale, the temperature is just right. If Earth were any closer to the Sun, water would boil; if it were further away, water would freeze. We need only to peer at Venus, our nearest planetary neighbour, to see the effect of straying from this cozy position. Being that bit closer to the warmth of the Sun, Venus has a runaway greenhouse effect, making it the hottest planet in the Solar System despite not being the innermost. So on their quest to answer the question of whether we are alone in the universe, astronomers have focused their attention on finding replicas of Earth beyond our Solar System.

FINDING A DARK NEEDLE IN A BRIGHT HAYSTACK

Detecting planets orbiting around other stars is tricky. For a start, stars are massive and planets are small in comparison – the Earth would fit inside our Sun over a million times. Not only that, but even the nearest star (Proxima Centauri) is hundreds of thousands of times further away than the Sun. Add in the fact that stars generate their own light but planets don't, and you are hunting for a very small, dark needle in a very distant, glaringly bright haystack. The challenge is so extreme that in almost all cases astronomers are reduced to inferring the presence of these extrasolar planets – or exoplanets – rather than observing them directly.

If things are lined up properly, as an exoplanet orbits its star it will move directly between the star and Earth, an event called a transit. In doing so, it will prevent a small amount of the star's light from reaching our telescopes and astronomers will spot a dip in the star's brightness. However, these dips are tiny – if aliens were watching Earth transit the Sun, they would only see our star's brightness alter by 0.01 per cent. Even Jupiter, which is much larger than Earth, would only block out 1 per cent. If these dips in brightness happen at regular intervals, they must represent each successive orbit of the planet and astronomers can be sure they've found an exoplanet. If the star is similar to our Sun and the dips in brightness occur roughly once a year, then they know that the planet must orbit at a similar distance to that of the Earth from our Sun and therefore that the exoplanet sits in its star's Goldilocks Zone. By the

Solar Dynamics Observatory (SDO) sees Venus (black) moving in front of the Sun during the Transit of Venus of June 2012. A similar event is used to detect distant planets by looking for drops in the brightness of the star as an exoplanet "transits" in front of it.

amount the star dims, astronomers can also tell how big the planet is. The Kepler space telescope has surveyed some 150,000 stars since it launched in 2009 and has used this technique to identify thousands of potential exoplanets.

There are other detection methods, though. The very first planet to be found orbiting another Sun-like star was 51 Pegasi b, which is about 50 light years (475 trillion kilometres, or 290 trillion miles) away and was named after Pegasus, the constellation where it resides in our night sky. Announced in 1995, this exoplanet was detected using a method known as the radial velocity technique. It is well known that stars pull on planets – that's why the Earth orbits the Sun – but a planet also exerts a tiny gravitational pull on its star. This causes the star to wobble slightly, in some cases making it move a little closer and then further away from the Earth. When the star is moving toward us, the starlight is squashed together, making it appear more blue. When the star is moving away, its light is stretched out, making it appear redder. Similar to the changes in pitch of an ambulance siren as it hurtles past you, these changes in the colour of the starlight are known as Doppler shifts. From the precise amount the light is shifted, astronomers can calculate how heavy the planet is – the heavier the planet, the bigger the wobble in its star and therefore the bigger the light shift.

However, finding an exoplanet roughly the same size as the Earth, orbiting it roughly the same distance as our planet, isn't enough – the planet probably needs water if it is going to be a cradle for life. Luckily, astronomers have a method for finding out which chemical elements are present on these distant worlds. When an exoplanet and its star are aligned in the right way, some of the starlight will pass through the planet's atmosphere and continue on its journey toward our telescopes. That passage through the planet's gas layers leaves telltale fingerprints in the light, betraying the gases they contain. Using an instrument called a spectrometer, astronomers can break the light up into the colours of the rainbow in a similar way to a prism. However, the starlight's spectrum will be missing particular colours, with dark bands where those colours should have been, because the light has been absorbed by the atoms in the exoplanet's atmosphere. So these missing colours reveal which elements are in the planet's atmosphere. This allows researchers to discover whether the exoplanet has water and oxygen, crucial ingredients for life here on Earth. The technique has been used on a handful of exoplanets – including GJ 1214b, a large, hot world 40 light-years (380 trillion kilometres, or 232 trillion miles) distant that is likely to be enveloped in a dense layer of steam – but a truly Earth-like planet has yet to be found.

But because a planet has the ingredients for life doesn't mean it is actually home to life. Biologists are still arguing over how life originally started on our own planet, and so it is not necessarily the case that, just

A small part of the Allen Telescope Array in California. The radio dishes – each around 6 metres (19 feet) in diameter – work together to listen to space in the hope of detecting signals from an alien civilization.

INSIDE EXPERT: WHY IS LOOKING FOR ALIEN LIFE IMPORTANT?

"A real buzz has been growing in astrobiology over recent years. The more we discover about the incredible survivability of life on Earth, the other environments in our own solar system and the abundance of exoplanets around other suns, the more likely it seems there could be life beyond our home world. If it's out there, there is a good chance that it is our generation who will find the first signs of alien biology, either in the dusty soil of our next-door neighbour Mars, or in the oxygen-rich air of an Earth-like exoplanet. The ultimate hope for some scientists is that we may even come across complex animal life forms – intelligent species that we could have a conversation with. But even if we only discover 'boring' bacteria on Mars, perhaps just fossils that have been long dead, it will revolutionize the way we see the world. We would know we are not alone in the universe, and that even life itself is commonplace. Such fundamental knowledge would be as profound as realizing that humanity evolved from ape-like ancestors, or that the Earth orbits the Sun. But the discovery of microbes living in our solar system would only be the first step. Biologists would then study and scrutinize this new example of life. We'll peer under its hood to see how it works, and compare how these cells are similar to our own. But, perhaps even more excitingly, we might find ways that they are truly *alien:* built in a fundamentally different way to us. Because I think by doing that we'll learn an enormous amount about ourselves, and our own place in this vast cosmos."

**Dr Lewis Dartnell, astrobiologist,
University College London, UK**

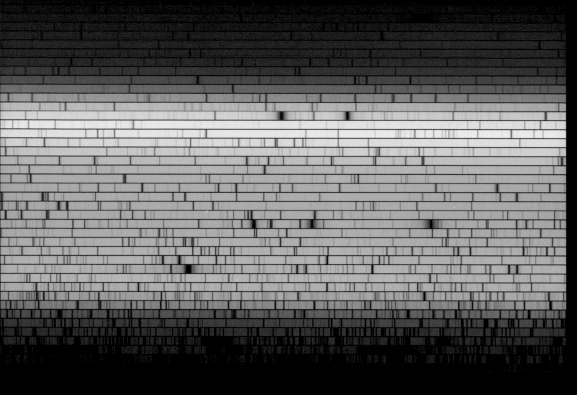

because the conditions are favourable, life actually got started somewhere else. Without travelling to these planets, which would take hundreds of thousands of years with current rocket technology, it is difficult to say any more than they might have life – a rather unsatisfactory answer to the question of whether we are alone in the universe. However, if there is life on another planet, and that life has evolved to exploit technology like we have, there might just be a way to prove they exist: by eavesdropping on them.

Spectrum of the Sun from the McMath-Pierce Solar Facility at Kitt Peak National Observatory, Arizona, USA. The dark bands act like fingerprints, revealing which chemical elements are present in our star. A similar technique is used to see if distant planets have water and oxygen.

COSMIC COMMUNICATION

The Earth is not a quiet planet. In fact we're pretty noisy – our radio and television signals, along with radiation from atomic weapons' testing, have been beaming out into space since the 1930s and 1940s. Travelling at the speed of light, they annually cover a distance of one light-year (about 9.5 trillion kilometres, or 5.8 trillion miles). That means in the intervening decades they have been travelling further and further away from Earth. Those first signals are now over 80 light-years (760 trillion kilometres, or 464 trillion miles) away and mark the boundary of the radio sphere – an ever-growing shell of radio waves with the Earth at the centre. Exoplanets have already been discovered that are not only within this radio sphere but also sit in the Goldilocks Zones of their stars. If intelligent life exists on any of these planets, it could be tuning into our signals. Equally, any intelligently inhabited exoplanet would likely have a radio sphere of its own and, if we sit within it, we'd be able to detect their signals too. Scientists involved in the search for extra-terrestrial intelligence (SETI) are dedicated to doing just that, using vast arrays of radio telescopes to listen in for ET's transmissions.

One observatory tasked with this work is the Allen Telescope Array, 470 kilometres (290 miles) north-east of San Francisco, California. Here, a series of radio dishes, each 6 metres (19.7 feet) across, stand side by side and work together listening for radio signals from space. It is hoped that future additions to the site could bolster the total number of dishes to 350. Astronomers have already pointed the telescopes at some of the potentially habitable planets that have been discovered by the Kepler space telescope in recent years but have yet to find radio signals that might indicate intelligent life.

But just as you have a choice of stations to tune your radio to, there is an enormous range of potential frequencies for astronomers to investigate. Fortunately, nature helps narrow down where we can listen. At frequencies

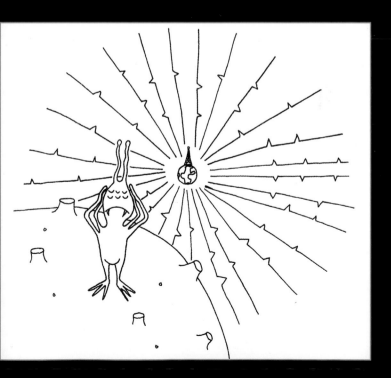

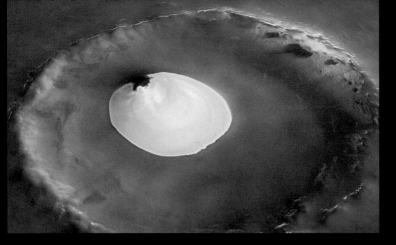

← - -

A Martian crater filled with water ice, imaged by the Mars Express spacecraft in 2005. In Mars's warmer past this would have been a lake, but today liquid water cannot flow freely on the Martian surface.

lower than about one billion hertz (1 gigahertz, or 1 GHz), any artificial signal would be drowned out by the noise of natural cosmic phenomena, such as pulsating stars and radio-emitting galaxies. At frequencies greater than 30 gigahertz, the background noise known as the cosmic microwave background – an echo of the radiation remaining from the Big Bang – is louder than any artificial signal would be. Couple this with the fact that water vapour in Earth's atmosphere prevents ground observations of frequencies around 20 gigahertz, and the observation window becomes even smaller.

Despite being narrowed by nature, this radio window is still huge and listening to all its possible frequencies is a Herculean task. In an attempt to make some progress, researchers have selected a sliver of this window bracketed by two of nature's loudest radio signals. The most abundant source of radio emissions in the universe is neutral hydrogen, which emits very precise radiation with a frequency of 1.420 gigahertz. Second only in strength to this are the 1.662 gigahertz emissions from hydroxyl compounds – chemicals containing one oxygen atom bonded to one hydrogen atom. Two hydrogen atoms bonded to one oxygen atom makes H_2O – or water – and the theory is that intelligent civilizations, almost certainly based on water, would choose to send their messages in the gap between the hydrogen and hydroxyl frequencies because they have the best chance of being heard. As US astronomer and one of the pioneers of SETI, Frank Drake, once said, "Let us meet, as different species have always met, at the waterhole!".

So whether it is exploring the planets and moons of the Solar System or planets orbiting other stars, the mantra of those trying to find our cosmic neighbours is "follow the water". Alien life might not necessarily need water, but we know water is so crucial to life here on Earth that it is an excellent place to start.

- - →

A 1977 print out from the Big Ear radio telescope at Ohio State University. Astronomer Jerry R. Ehman scribbled Wow! in red ink because the signal bore the hallmarks of being artificial. It remains the only signal detected to date that could have come from alien life.

BIG DISCOVERY:
THE WOW! SIGNAL

In August 1977, US astronomer Jerry R. Ehman couldn't believe what he was seeing as he scanned through the recent printouts from an Ohio State University radio telescope affectionately called "Big Ear". Such was his disbelief that he scribbled "Wow!" on the printout in red ink. What the telescope had picked up was a very loud, 72-second-long blast of focused radio waves that seemed to be coming not just from beyond the Earth but from outside the Solar System. What's more, the frequency of the signal was very close to 1.420 gigahertz – the approximate frequency astronomers guess aliens might prefer to communicate at. Could someone have been trying to get our attention?

Ehman and colleagues set to work trying to trace the source of this strong signal, ruling out a whole host of alternatives, and the actual source of this tantalizing emission remains a mystery to this day. If aliens were trying to contact us, you might expect them to periodically repeat the signal, but despite attempts to pick it up again in the 1980s and 1990s, the signal has never been heard again. It could be that we have already eavesdropped on our galactic cousins, but without a repeat signal it can never be conclusively proved. More than 35 years on, the Wow! signal remains the only recorded radio signal that has the hallmarks of being artificial, but it is unlikely we'll ever know for sure.

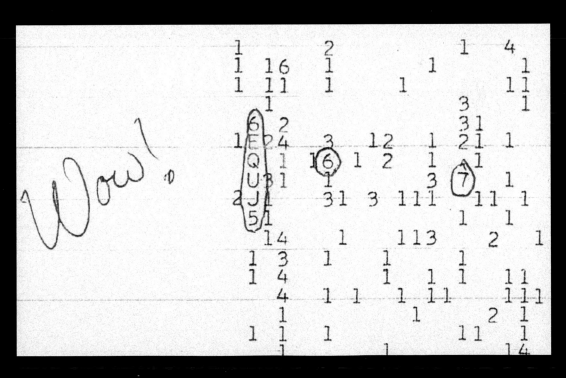

4

WHAT MAKES US HUMAN?

Kanzi is a thirty-something bonobo living a nice life at the Great Ape Trust/Bonobo Hope Sanctuary in Des Moines. He does what most bonobos do: laze around, play and eat fruit. But he also fashions and wields a variety of flint tools, much as we imagine our ancient ancestors did. Left with a log with food hidden in it, Kanzi uses smaller flints like a drill bit to scrape the seams, larger ones like an axe and wedge to split the log apart, and a flat oval flake as a spade to scoop the food out. He also knows how to dig for food in different soils – hands are fine for soft sand, but muddy soil needs branches and hard soil needs more solid stone tools. And if that hasn't impressed you enough, he also knows sign language.

Toolmaking is among the large list of traits we used to think were uniquely human, like walking on two legs, morals and humour. But over the years we've had to cross off larger and larger portions from that list. So proud were we of our tool prowess that we even named one of our ancestors after the trait (*Homo habilis*, which literally means "handyman"). Kanzi is just one creature who's shown us this. We've even seen other apes hunt with spears; capuchin monkeys use stones as hammers to break nuts; and vultures use pebbles to break eggs. Even octopuses have been seen collecting and transporting coconut shells for use as portable shelters. And that's just tools. We now know elephants work together, parrots can be pretty good at maths, dolphins can distinguish mirror-images of themselves and even bees can recognize faces.

Yet the fact that there are seven billion of us humans dominating the planet, the quest to understand why it's not a planet of the apes or dolphins or any other kind of animal endures today. The father of evolution, Charles Darwin, believed "the difference in mind between man and the higher animals, great as it is, certainly is one of degree and not of kind". But even he would have been astounded at just how narrow that degree of difference has become.

The human genome is 99 per cent similar to a chimpanzee's and, for that matter, 50 per cent similar to a banana's. Even if we look at just the genes that code for functioning proteins, we are 69 per cent similar to yeast and 44 per cent similar to fruit flies. Narrowing the focus even further to the genes at work in the brain, we still see very little difference between chimpanzees, humans and even mice. We had great hopes that sequencing the DNA of monkeys, chimpanzees, orang-utans and the caveman Neanderthal would reveal significant differences between us and them, but these have not turned out to be as dramatic as we'd hoped.

We should, perhaps, have expected this. Scientists already had a shock when the first draft of the human genome was published in 2001. It had been commonly assumed that humans must have more genes than other species – some estimated as many as 100,000. The finished sequence revealed the actual figure to be much lower, even as low as 20,000. Our

Kanzi, the bonobo chimp who knows sign language.

← – –

Sir John Sulston, one of the pioneers in the Human Genome Project. He won the 2002 Nobel Prize in physiology or medicine for his efforts.

↑
│
│

the years, we've come to terms with why. Studies have disproved the idea that any DNA outside of the genes – the segments of DNA that carry our traits from parent to offspring – is "junk". We now know that this "junk" DNA often codes for important elements that regulate the genes – how, where, when and how much they're expressed. This could account for differences in development, function and behaviour from species to species. For example, humans produce more copies of a protein called DUF1220 than other species (272, which is twice as many as chimpanzees and 272 times more than mice and rats, who have only 1). Indications are that such diffferences in gene expession may have helped with our large brains (more on which later).

However, it's also clear that losing, as well as gaining, chunks of DNA has made us who we are. Studies suggest that we've lost 510 bits of regulatory DNA found in other species, such as mice, chickens and chimpanzees. These lost bits of DNA could have been involved in all kinds of things – possibly, for example, the spiky penises with barbs made of keratin (the stuff in hair and fingernails) found in other male mammals but lost from the human lineage sometime after we diverged from chimpanzees 5–7 million years ago.

When looking at such changes, the difficulty comes in working out exactly what each change did, particularly when it's in a regulatory element several stages removed from the actual gene that is ultimately affected. The size of the change is not much help, either – large changes to the DNA may make no difference while small changes may have huge effects. This also makes it difficult to distinguish between changes that were significant and effective enough to have been selected and conserved through evolution and those that occurred just as naturally but had no effect on our body or behaviour.

It's no easier moving back up from the regulatory DNA to the level of genes, the carriers of hereditary traits. We have been able to identify some important genes that natural selection has kept since we branched out from the rest of the apes. For example, the genes ASPM and MCPH1, which are associated with brain size, and FOXP2, which has an important role in human speech. However, the human version of FOXP2 differs only slightly from the chimpanzee version, and the gene is also common in many other species. Even when more distinct genes are found, we're still not sure what many of them do.

Moreover, each evolved function, even if significant, doesn't in itself necessarily define humanity – we know we've evolved to combat malaria, but that by itself doesn't make us human.

Printouts of the human genome in the map room of a laboratory near Paris.

A BIG BRAIN, BUT DOES SIZE MATTER?

The human brain weighs a massive 1.3 kilograms (about 3 pounds), which puts us among the heavyweights of the animal kingdom when it comes to brain size, although we do not have the biggest – whales, elephants and even dolphins have bigger, heavier brains, as did the Neanderthals. Instead, where our brains pack a punch is in the number of nerve cells: there are an astounding 86 billion neurons in the human brain – a thousand times more than in a mouse's brain, about five times more than in a baboon's and three times the number in a gorilla's. It's a huge number and one that many think makes a difference, even if we're still working out exactly how (*see* QUESTION 5: WHAT IS CONSCIOUSNESS?).

One theory was that among those billions of neurons, there might be specific types unique to humans. If this sounds a bit like wishful thinking, you'd be right. Spindle neurons are common in brain areas linked to emotional judgements and social interactions, and we once thought these neurons were unique to human brains. However, scientists have found that whales have three times as many in the same brain areas and, evolutionarily speaking, they've probably had them for twice as long as we have. Spindle neurons have also been found in dolphins and elephants, and although they're likely to have evolved these neurons independently rather than from the same ancestor as us, it doesn't make ours any more special.

It's also been suggested that our growth in brain size boosted the size and power of key brain areas. Certainly, looking at the skulls of Neanderthals we can see how our ancestors' brains were possibly more developed than theirs in two areas: the temporal lobe, associated with hearing, language

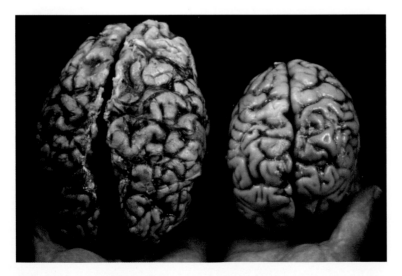

A human brain (left) and a gorilla brain (right). Ours are not only significantly bigger but contain three times as many neurons.

← - -

and speech, and the parietal lobe, involved in spatial awareness. Both these areas are in the wrinkly, outer bit of the brain known as the cortex, but other parts of the cortex aren't that different; for instance, the human frontal cortex, which is associated with problem-solving and intelligence, is proportionally a similar size to that of other mammals.

What triggered our big brains is also still up for debate. Sometime 1.5–2 million years ago, our ancestor *Homo erectus* ("upright man") began something of a growth spurt in his mental capacity. Some researchers have suggested that chance genetic mutations (which occurred in humans but not chimps) gave us weaker jaw muscles that helped to reshape the skull, enabling it to accommodate a larger brain. Others have proposed that mutations created the nerve signalling pathways that, refined over thousands of years of evolution, shaped the brain we have today. Another theory makes a strong case for the increase in brain size being driven not by genetics, but by cooking. There's only so much nutrition the human digestive system can extract from raw food (an estimated 30–40 per cent of the food's total nutritional content). Cooking breaks down cell barriers and releases nutrients we otherwise wouldn't be able to access,

allowing us to extract almost all the nutrients. As you can imagine, the billions of neurons in the brain require a lot of energy to keep going. In fact, scientists estimate that 20–30 per cent of our entire body's energy output goes to the brain. It is thought that our ancestors' shift to cooked food gave them more calories, which helped to fuel their growth in brain size. This, in turn, has led to speculation that our mastery of fire is what propelled humans to global domination. It's likely that a combination of all these factors gave us our big brains, but whichever ones turn out to be true, the result is the same: our higher brain capacity allowed us to develop the skills, technology and cognitive behaviours that set us apart from the Neanderthals and the apes.

Language, the ability to articulate our thoughts, is considered by many to be one of the defining features of humanity.

APES AND HUMANS – NOT SO DIFFERENT?

The cognitive similarities and differences between us and apes was shown in an interesting series of experiments by Michael Tomasello and his colleagues at the Max Planck Institute for Evolutionary Anthropology, Germany. They employed a bunch of cognitive tests normally used to assess primates but also applied them to two-year-old children. Surprisingly, chimpanzees and orang-utans proved just as capable as the toddlers when it came to tests of "physical" intelligence – locating a reward hidden under a cup (food for the primates, a toy for the toddlers), understanding how a tool could be used to get the reward and dealing with different quantities. The children did, however, perform twice as well as the chimpanzees or orang-utans in tests of "social" intelligence – copying an adult's actions or noticing their social cues to get the reward. So in our early cognitive development we may start off at a similar level to our closest simian cousins, but by the age of two we're already outperforming them in other ways.

One of those ways is language. This, giving us the ability to articulate our thoughts, is often thought to be one of the defining features of humanity. The US linguist Noam Chomsky proposed that you need language before you can have thought. His theory is that all human languages are similar and are based on our mind's innate capacity for language. But the uniquely human aspect of language is, according to Chomsky, "recursion" – the ability to insert one phrase into another of the same type and so combine discrete thoughts, as in combining "I am reading this book" and "I am sitting down" to produce "I am sitting down, reading this book". Michael Corballis, a psychologist at the University of Auckland, New Zealand, has suggested that recursive thought allows us to engage in a sort of mental time travel, whereby we can recall past episodes, imagine future ones and sometimes even insert fiction, all of which can be useful in planning and predicting. It's a compelling argument. The problem is that there's some evidence

INSIDE EXPERT: IS WHAT MAKES US HUMAN ONLY IN OUR GENES?

"The populist statement that the genome is a 'blueprint' to make a human is incorrect. The genome is more like a recipe book. So the environment in which you're cooking makes a big difference, as does who is doing the cooking. Genetics puts some constraints on what you can become, but it doesn't determine exactly who you are.

For instance, much of human cognitive and physical development is postnatal – it happens after birth. We have one of the longest postnatal development phases of all species – the brain can still be undergoing development until we're over 25. We also seem to be the only species with a long adolescent phase, as opposed to a short 'juvenile' phase, between childhood and adulthood. This means that the environment can potentially make a much bigger difference to our development than it could to a rat or mouse.

The important thing is cognitive development. Something appears to have happened 100–200,000 years ago that allowed *Homo sapiens* to emerge from among the other human-like creatures there were around at the time. We're still working out what it was, but the cognitive abilities we gained probably allowed our ancestors to do what many other species couldn't, and we proceeded to replace them.

When we sequenced the chimpanzee genome and compared it to ours, some thought this would reveal the differences that made us human. In retrospect, this was more a hope than an expectation. With millions of years of evolution since our common ancestor with the chimpanzee, it was never likely to be single gene that made a difference, but a combination of hundreds or even thousands of genes and genetic changes, interacting with the environment. And it is perhaps the resulting transition in cognition that separated us from the Neanderthals and others."

Ajit Varki, Co-Director, University of California San Diego/Salk Institute Center for Academic Research and Training in Anthropogeny (CARTA), USA

Our ancestor *Homo habilus* (literally "handyman") who lived approximately 2.3 to 1.4 million years ago and used tools.

that songbirds may do it, and we're not even sure all of us humans do it. Daniel Everett, another US linguist, has spent most of his life studying the Pirahã tribe of the Amazon and sees no evidence of recursion in their language. They also don't appear to use numbers, words for colour, any quantifying terms (such as "few" or "many"), the present tense, deep memory or drawing. This almost complete lack of conventionality makes their culture fascinating, but it also raises questions about whether our language is indeed what makes us special.

CULTURE VULTURES

It seems increasingly likely that it's not just physiological or genetic factors that make us human, but culture too. Various scientific luminaries, from Charles Darwin to the British evolutionary biologist Richard Dawkins, have alluded to the importance of culture and how it crosses over with genetics. The theory goes that genetics and culture actually reinforce each other. Scientists have proposed mathematical models of how genetics can help favour the selection of cultural traits and how they, in turn, can hasten the selection of genetic change. If you like cheese or your tea milky, you're demonstrating a prime example of this. Like most mammals, we rely on milk as a source of sustenance as babies. But unlike most mammals, we go on drinking it all the way through childhood and into adulthood. The key is the enzyme lactase that digests the sugar lactose in milk. In most humans, the lactase gene continues to be active, instead of switching off after weaning as it does in most other mammals. We think that the cause of this may have been the adoption of dairy products in the diet in certain cultures, which made it evolutionarily beneficial to have the lactase gene active for longer. The evidence for this idea is that mutations in the lactase gene that keep it active have arisen independently in different population groups around the world, possibly linked to the domestication of cattle. In contrast, groups that don't traditionally include dairy products in their diet – many African and Asian populations, for instance – tend not to have these activity-prolonging mutations.

But perhaps the key to humanity's uniqueness lies in economics, specifically, in trade. The law of competitive advantage roughly says that we all gain by trading with one another as long as each of us is more efficient at making different things. Let's say I'm better at making spears whereas you're the best axe-maker in town. Rather than each of us making one good thing and another of poorer quality, if I make two really good spears and you make two really good axes we can trade and both win. Put that on a twenty-first century economic scale and you can see how it benefits society – I don't farm all day for

Humans are unusual in drinking milk long after infanthood. This cultural habit has affected our genetics, selecting for an active lactase gene, which is turned off in most mammals after weaning.

food, but working for an electricity company I might provide power for the farmers who do. This kind of simultaneous reciprocal trade in skills and products seems to be peculiarly human. Chimpanzees can be taught to trade to some extent, but they won't hand over food they like for food they like more. It may have started with the sexual division of labour seen in hunter-gatherer communities, with males hunting and females gathering. However it began, it might be the reason we've been able to make huge strides as a species – by helping each other out. As the British science writer Matt Ridley puts it, "Human achievement is based on collective intelligence – the nodes in the human neural network are people themselves. By each doing one thing and getting good at it, then sharing and combining the results through exchange, people become capable of doing things they do not even understand". So perhaps what makes us human is not a gene or a brain or anything else individual, but humanity as a whole.

5

WHAT IS
CONSCIOUSNESS?

I magine for a moment you are not dead, but not asleep, not in a coma nor in any way aware of your surroundings. In fact, you are not aware of anything. You feel no pleasure or pain, or any bodily sensation of any kind. You have no thoughts, memories or emotions, nor are you able to follow or understand speech. You still have basic brain functions, such as those that control breathing, but how much of your higher, cognitive functions remain is a mystery. This condition is known as a vegetative state and is one of the most horrible potential consequences of brain damage. Though there may be occasional signs of life – opening the eyes, moaning, smiling or crying – more often than not these are just random responses. Sadly, few recover from a vegetative state. But what if a person isn't actually in a vegetative state and is still conscious?

In a brilliant series of experiments, a team led by Adrian Owen, then at the University of Cambridge, enabled a 23-year-old woman who had been tragically incapacitated by a traffic accident to prove not only that she maintained some level of consciousness but also to communicate, simply, with the outside world just by thinking about tennis. The key to this is the knowledge that the pattern of brain activity involved in playing tennis is very different to that involved in walking around your home. Imagining tennis activates a part of the brain involved in planning movements – an area known as the supplementary motor cortex – whereas walking around requires another area of the brain – the parahippocampal gyrus – for spatial navigation. Armed with this knowledge, Owen's team played the woman pre-recorded messages, one asking her to imagine playing tennis and the other walking around her home, while her brain was being scanned. Remarkably, the scans showed a clear difference in her brain activity in response to the two messages, and the patterns of her brain activity were similar to those of healthy volunteers. What's more, the patterns lasted more than 30 seconds until she was asked to stop, which the researchers believe to be a clear demonstration of conscious motivation rather than a transient, automatic response. Owen then took things a step further, using a similar method to ask simple yes/no questions of 54 people who were minimally conscious or in a vegetative state. Five of the 54 responded correctly (as indicated by their brain scan patterns), sometimes answering questions that only they or close friends and family members would have known. All five had been diagnosed as being in a vegetative state.

These studies raise the possibility of using this and other techniques not only to better diagnose minimally conscious patients (Owen estimates that close to 20 per cent of patients thought to be in a vegetative state are actually conscious but incapable of proving it through standard clinical tests), but also to ask them simple decision-making questions, perhaps about treatment, such as painkillers, or their emotional state. Or even to ask if they wish to remain on life-support, although this question is loaded with all sorts of ethical, legal and technical issues concerning

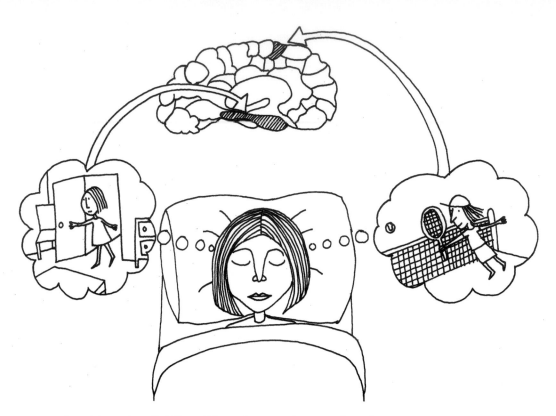

among other things, the reliability of the technique, the person's quality of life and just how conscious their conscious state is. As Owen himself has said, "We can't get inside [the patient's] head and see what the quality of her experience is like." Therein lies the heart of the question of consciousness: how do we get inside our minds, our psyche, our "self", and what is there?

The human brain weighs around 1.3 kilograms (about 3 pounds) and contains some 86 billion neurons – a thousand times more than that of a mouse. These are impressive statistics but don't explain why the brain is conscious, what consciousness is or even what exactly consciousness means. For consciousness, like many things in the non-physical realm of the psyche, is intangible, defies definition and, as people in a vegetative state show, is difficult to prove. Indeed, in 2004 a group of neuroscientists refused to define it, writing: "[Humankind has] no idea how consciousness emerges from the physical activity of the brain and we do not know whether consciousness can emerge from non-biological systems, such as computers... Currently we all use the term 'consciousness' in many different and often ambiguous ways...[it] has not yet become a scientific term that can be defined in this way."

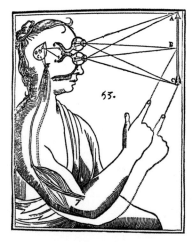

Putting aside the fact that we're not entirely sure what it is we're discussing, the thinking goes that if we find out which bits of the

A diagram of the mind from René Descartes' works.

brain are involved and how the neural circuitry works, we'll figure out how consciousness emerges. So the first question of consciousness is: where is it located? The ancient Greeks thought it was in the heart, the Mayans believed it was in the liver, and the seventeenth century French philosopher René Descartes thought it was in the pineal gland (a tiny, cone-shaped gland in the middle of the brain). However, we've gradually come to realize that there is no single part of the brain responsible for consciousness. Instead, it arises more as a result of a collective of brain areas working together to different degrees and in different capacities. By analysing brain activity during conscious and non-conscious thought, we've found several parts of the cortex (the outer, wrinkly part of the brain) to be involved, forming a network with the frontal lobe (involved in attention, planning and language) at the front of the brain, and the parietal, occipital and temporal lobes (processing sensory information, vision and sound, language and memory) at the rear and sides of the brain. Also playing a key role is the thalamus, a large, two-lobed structure buried under the cortex that acts like a gatekeeper, determining which nerve signals to pass on to the cortex. However, the thalamus is more than just a relay station, it also helps to control the level of consciousness; indeed, damage – specifically, to a tiny area called the centromedian nucleus – can cause a person to lose consciousness, in some cases into a permanent coma. Nasty as brain damage is, it has taught us a lot about how different parts of the brain work together, with knock-on benefits for the study of consciousness. Corpus callosotomy is an extreme but sometimes necessary surgical treatment for severe epilepsy in which a surgeon severs the bundle of nerves connecting the left and right sides of the brain. Incredibly, this not only separates the two halves of the brain, but also the patient's consciousness.

Pioneering (and Nobel-Prize winning) experiments by US neurobiologist Roger W. Sperry in the 1960s showed just how unaware of each other the two sides of a split brain are. The brilliance of Sperry's experiments on split-brain subjects lies in their design – a screen flashing images only to the left or right visual field – and two important facts: the right side of your brain controls and receives sensory signals from the left side of your body (and vice versa), and speech is controlled by areas found only in the left side of the brain. So since nerve signals from the right visual field go to the left side of the brain, where the speech areas are located, split-brain subjects have no problem naming objects shown on their right. But when images are shown only to the left visual field, which sends nerve signals to the right side of the brain, the subjects are unable to access the words, and answer, "I see nothing". Things become even stranger when a subject is asked to touch an object hidden behind a screen – say, a fork – with the left hand and then describe it. As with the visual fields, nerve signals from the left hand go to the right side of the brain, which knows it's a fork but the subject can neither say the word nor can they say "nothing", because they are clearly holding something. Faced with this dilemma, the left side of the brain takes a guess: it's a spanner.

Another bizarre phenomenon that can occur in split-brain patients is alien hand syndrome, in which one hand seems to have a mind of its own. This, we think, results from the fact that one side of the brain is more dominant than the other (think right-handedness). Without instructions from the dominant side due to the split brain, the other side lacks instructions and attempts to join in the activity or improvise. The result: an "anarchic" hand that does things like unbuttoning the shirt you've just done up with the other hand.

So we know that networking of brain areas is important in consciousness, but many mysteries still remain. For example, how is it that we can lose large chunks of the brain – including the cerebellum, which contains more than half of the brain's neurons – without affecting consciousness? The fact is we still don't know enough about how all the areas of the brain work together and how that determines conscious experiences or, indeed, even the level of consciousness

THE UNCONSCIOUS BRAIN

Consciousness can be viewed as our general level of awareness or awakeness, particularly in a clinical sense. The scale goes from wide awake through inattentive and sleepy, all the way to brain dead. When we're asleep, we disengage from our sense of space and time but we certainly don't shut down completely (as was wrongly assumed until the 1950s). The brain stays conscious, going through cycles of relaxation and activity, and at some stages we are more "dead to the world" than others. And throughout, of course, we dream (*see* QUESTION 6: WHY DO WE DREAM?).

Besides sleep, there is another form of unconscious oblivion that we commonly, and voluntarily, enter: anaesthesia. Although anaesthetics have been used in medicine since the mid-nineteenth century, we still don't know exactly what happens when a patient "goes under" in general anaesthesia. Indeed, one in every 1,000–2,000 patients given a general anaesthetic temporarily regains consciousness or remains conscious during surgery. What we do know is that general anaesthetics affect the cortex and centromedian nucleus of the thalamus, which fits in with our theories about where consciousness is located. Rather than shutting down the whole brain, general anaesthetics somehow depress the electrical activity of the neurons, disrupting the network required for consciousness and possibly also the integration of information that comes from different parts of the brain, which some scientists believe is the very essence of consciousness itself.

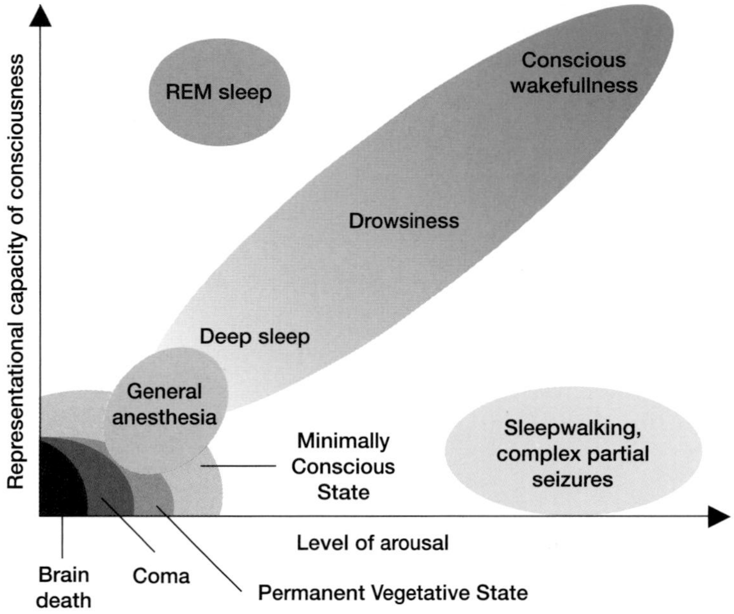

A graph illustrating the spectrum of known conscious and unconscious states. There are various states of 'wakefulness' and 'attentiveness', several of which remain a mystery.

Anaesthesia is just one of the unconscious states we still know little about but there are several others. Coma, caused by severe trauma to the brain's sensitive environment – a blow to the head or a change in the glucose or oxygen supply to the brain – seems a lot like deep sleep but doesn't have the typical sleep-wake cycles. A person in a coma has no sense of pain and gives no indication that they are aware of their surroundings. In addition to coma, there are further levels of unconsciousness in trauma patients, such as a minimally conscious state (MCS, a condition of severely altered consciousness, where there is minimal evidence of some form of awareness), persistent vegetative state (PVS), and locked-in syndrome (total paralysis of all the muscles rendering a person unable to speak or move). And this brings us back to our ethical dilemma: how can we tell what the conscious state of a person in a coma or vegetative state is like when we don't even know what a healthy person's conscious state is like?

THE HARD PROBLEM

The difficulty of knowing a person's conscious state is one of the "hard" problems, as the Australian philosopher David Chalmers put it. In a landmark 1996 paper, he wrote, "There is not just one problem of consciousness. 'Consciousness' is an ambiguous term, referring to many different phenomena. Each of these phenomena needs to be explained, but some are easier to explain than others." (*see* KEY IDEA: CHALMERS' EASY PROBLEMS OF CONSCIOUSNESS). His point was that there are problems that

INSIDE EXPERT: HOW CAN WE DEFINE CONSCIOUSNESS?

"I don't think it's possible to provide a useful, non-circular definition of consciousness. There are various synonyms for 'consciousness', such as 'awareness' and 'experience', but both of these terms are as difficult to pin down as 'consciousness' is. In my view, the best that one can do is to say that an organism is conscious just in case there is something that it is like to *be* that organism, as the philosopher Thomas Nagel has proposed. The phrase 'what it's like' isn't to be understood in a comparative sense, but is meant to point to the fact that consciousness involves having a subjective perspective or point of view on the world. One distinction that is useful in thinking about the scientific study of consciousness is the distinction between conscious creatures, conscious modes (or 'levels'), and conscious 'contents'. The first is straightforward: some creatures are conscious and some creatures are not. But conscious creatures also differ in the ways in which they are conscious. Some of these ways concern global aspects of consciousness – what are sometimes called 'levels of consciousness' or simply 'states of consciousness'. For example, there is the normal waking state, the REM dream state, the minimally conscious state, the state of light anesthesia, and so on. Another way in which consciousness is modulated involves fine-grained conscious states or 'contents'. For example, we enjoy experiences in different perceptual modalities – such as vision and audition – and within each perceptual modality there are further distinctions in the ways in which a creature can be conscious. So, to give a full characterization of a creature's consciousness, one must specify both its mode of consciousness and what contents it has in consciousness at the time in question."

Tim Bayne, Professor in Philosophy, University of Manchester

Rodin's *The Thinker* sculpture at the Rodin Museum in Paris

relate to awareness and how that works through systems, mechanisms or computation. These we are confident we can solve empirically through science, even if it might take some time to work out the details). The other aspect of consciousness – experience – is harder to explain. To paraphrase US philosopher Thomas Nagel, you can never be sure what it's like to be a bat, because you're not a bat. In other words, consciousness is a purely subjective, internal experience and therefore one being can never truly know what another being's consciousness is like.

Consciousness is not only difficult to pin down, it's also difficult to study because, as German philosopher Thomas Metzinger put it, consciousness is "thin" and "evasive". Much of what we sense, and many of our actions and bodily functions, take place under the unconscious control of a part of the nervous system called the autonomic nervous system. As a result, we have little access to what we're actually doing at any one time, but we think we know what we're doing. A perfect example of this is choice blindness. In one ingenious experiment, scientists asked customers in a supermarket to taste two very different flavours of jam – cinnamon-apple and bitter grapefruit – and say which they preferred. They were then asked to taste their preferred choice again and explain why they preferred it. However, unknown to the customers, the experimenters gave them instead their other, non-preferred jam. Surprisingly, less than a third of the 180 customers noticed the switch and instead justified their non-preferred choice. Choice blindness highlights the fact that our memories and viewpoints are notoriously subjective. We rewrite history after it happens, so we can never be sure when comparing experiences.

Nevertheless, experience is an essential part of consciousness, and that poses the real question: why? Why should such a thing emerge from a network of brain cells, and why would we have evolved it, and kept it? This, said David Chalmers, is the really hard problem: "Why is it that when our cognitive systems engage in visual and auditory information-processing, we have visual or auditory experience: the quality of deep blue, the sensation of middle C? Why should physical processing give rise to a rich inner life at all? It seems objectively unreasonable that it should, and yet it does."

We do have a few ideas, of course. A good suggestion is that integrating and processing lots of different information and neural activities together enables us to act quickly, which would have been very useful in the dangerous, dog-eat-dog world of our ancestors. Conversely, conscious thought allows us to focus, blocking out rather than reacting to the multitude of sensory inputs bombarding us every second and choosing what is most relevant or necessary. Combined, these two abilities could allow us to distinguish between what's real and what's not while also enabling us to think through multiple future scenarios and decide which actions might help us achieve, or avoid, them.

THE NATURE OF THE CONSCIOUS MIND

Through years of thought and research on consciousness, neuroscientists, psychologists, philosophers, computer scientists and others have tried to define what a conscious being is. This boils down roughly to five generally accepted mental attributes or abilities you must possess:

A sense of self – an awareness your own existence.

A sense of surroundings – a perception of yourself, your environment, your position in the environment and how you interact with the environment.

Subjectivity – your own, personal view of the world and yourself.

Sentience – the ability to sense your environment, in our case including through our main senses of touch, smell, taste, sight and hearing.

Sapience – the ability to have thoughts or feelings and act upon them in a knowledgeable way.

Others might also include imagination, which can encompass the ability to recall memories and combine them to represent future scenarios, making it possible to plan ahead; the ability to focus on one thing over another; and the use of emotions as a guide to what is good and bad for you.

These criteria are useful not just in helping us define what our consciousness is, but also whether we're alone in possessing it – and for many of the criteria listed above, the evidence indicates otherwise. Elephants can work together to solve problems; dolphins, magpies and various great apes can recognize themselves (and their behinds) in a mirror; and bees, which have a minuscule number of neurons compared to ourselves, can recognize individual faces. Even the octopus navigates using landmarks and spatial awareness and has been seen to plan ahead – collecting and transporting coconut shells for future use as a portable shelter.

Beyond simple behaviour observations, we see similarities in the brain circuits of mammals and birds. The cerebral cortex, for instance, is pretty consistent in its tiny size and shape throughout many species. Deeper studies of various animal brains have seen zebra finch birds demonstrate similar neural sleep patterns to mammals. Of course, as the "hard problem" states, we can't be sure what an animal is thinking or feeling. Nevertheless, a prominent group of scientists declared in their 2012 Cambridge Declaration of Consciousness that: "Convergent evidence indicates that non-human animals have the neuroanatomical, neurochemical, and neurophysiological substrates of conscious states along with the capacity to exhibit intentional behaviors. Consequently, the weight of evidence indicates that humans are not unique in possessing the neurological substrates that generate consciousness." The main difference, they say, is that we have the language to talk about it. As the scientists put it: "Until animals have their own storytellers, humans will always have the most glorious part of the story". After all, babies can't communicate either, but we wouldn't deny that they are conscious simply because their brains aren't fully developed yet.

What we would consider conscious or not also helps us in the ultimate quest that may bring us closer to understanding consciousness: to create it ourselves. The dream of artificial intelligence has been pursued in earnest since the 1950s (*see* QUESTION 13: WHEN CAN I HAVE A ROBOT BUTLER?), and robotics has a history of testing ideas on brain connectivity. In 1948, the neuropsychologist William Grey Walter developed simple

Octopuses have been observed navigating via landmarks and collecting coconut shells for future use as portable shelters. Such spatial awareness and forward planning is thought to be one of the hallmarks of consciousness.

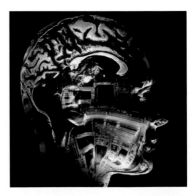

The quest to create an artificial intelligence may bring us closer to our understanding of consciousness itself.

KEY IDEA:
THE EASY AND HARD PROBLEMS OF CONSCIOUSNESS

"There is not just one problem of consciousness. 'Consciousness' is an ambiguous term, referring to many different phenomena. Each of these phenomena needs to be explained, but some are easier to explain than others."

The biggest problem with studying consciousness is knowing what exactly it is you are studying. To Australian philosopher David Chalmers, when it comes to consciousness, there are "easy" aspects and there are "hard" ones. The "easy" ones can, with time, probably be solved empirically through scientific observation and experimentation. They are "easy" because all that is required to solve them is to find a mechanism (no matter how complex) that performs the function. These might include:

- How we are able to distinguish, make sense of and react to sensory information from our environment;
- How we are able to integrate all that information;
- How we can access not just external sensory information, but information about our internal mental state, and describe it through communication;
- How we are able to focus our attention on one thing (or several things) over others;
- How we deliberately control our own behaviour;
- What the difference is between being "awake" and "asleep", what lies in between and what we can and can't do in these various states.

Of course, these are anything but easy to answer, but they certainly seem so in comparison to the "hard" problems of: How to measure conscious experience, and be sure of it, when it is by its nature purely subjective, *and* why on earth consciousness should arise in the first place?

Chalmers' definitions didn't answer the question of what consciousness is, but by breaking it down he provides a slightly clearer framework for the question, distinguishing the aspects that we can investigate from those that require a lot more thought.

robotic "tortoises" to represent brain cells and demonstrate how their interactions could lead to rich behaviours. Today's approaches to artificial intelligence are fascinatingly sophisticated, from training a "child-like" program through its own trial-and-error experience or human teaching to developing a robot capable of forming an "internal model" of itself, its surroundings, other robots and maybe, one day, humans. The machines are getting closer. The dream remains alive, fuelled by the idea that if we can understand and replicate the workings of the human brain, consciousness will emerge.

WHY DO WE DREAM?

If you were offered a pill that could keep you permanently awake, would you take it? Would you be tempted, knowing that you could achieve so many more things in 24 hours without ever having to feel tired? Or would you crave the simple pleasure of a good night's sleep? For many of us, those few precious hours in bed may be the only respite we get from our otherwise hectic lives. Our brains are in overdrive from morning till night, helping us to deal with everything we are confronted with on a daily basis – from negotiating traffic, to punching in PIN numbers and locating lost door keys. With so much going on, it is no wonder we long to crawl into bed at night, seeking solace for our busy brains. But while sleeping gives our bodies a chance to recharge, science tells us that our minds are far from at rest. Ever since it has been possible to monitor the brain waves of sleeping volunteers, it has been obvious that during slumber, the brain is up to something – something that we are not fully aware of. Dreaming gives us a window on the brain that might allow us to work out what is going on during these nightly activities. More than this, though, dreaming itself may play an essential role in moulding our experiences of the world and helping us to make sense of everything in it. Unfortunately, from a scientific perspective, the whole concept of studying dreams is problematic. Science, which prides itself on its accurate measurements and objective observations, is altogether at odds with a subject that is necessarily based on second-hand reports of events that occur only in the mind. It is perhaps for this reason that until the nineteenth century there was no scientific experimentation in the area of dreaming. Until this time, a dream was widely considered to be a divine message or a prophecy about the dreamer's future.

It was from an unlikely background that one of the first real scholars of dream studies emerged – Frenchman Louis Ferdinand Alfred Maury was a historian and archaeologist, but the skills he gained as an archivist gave him the approach that was needed to start looking at dreams in a more scientific way. Maury kept a diary of his dreams for no less than 20 years and carried out what are considered to be some of the first cause-and-effect experiments on dreaming, in which he not only designed the experiments but also acted as subject. He never revealed the identity of his assistant, so we must assume that he asked his wife or a long-suffering servant to help test the effects that different disturbances would have on his dreams, which he faithfully recorded in his journal. Among these effects, for instance, he noted that after having his nose tickled with a feather he dreamed about wearing a mask that pulled off his skin, and the smell of perfume conjured up dreams of Cairo. However, of all the strange dreams Maury described in his journal, his best known is referred to as "the guillotine dream", made famous by Sigmund Freud in his book *The Interpretation of Dreams*. In this dream, Maury found himself on trial for political offences during the French Revolution and condemned to death by guillotine. His account states that he felt his head separate from his body as the blade dropped. Maury then awoke to find that his bedpost had broken off, hitting him on the neck as it fell. This caused some debate in the literature, since others contended that such a dream was too long to have been triggered by the bedpost. It also seems that Freud embellished this story to suit his theory that dreams were expressions of unfulfilled wishes – in his retelling of it he included a part where Maury kissed a woman farewell before boldly stepping up to the platform. Whatever the truth of the matter, Maury had begun a trend in documenting dreams, inspiring many nineteenth-century doctors and psychologists to discover what they could learn from their own dreams.

INVESTIGATING THE INTANGIBLE

For well over a century now, scientists have relied on the equivalent of a pen and paper beside the bed to document dreams. Since even the most sophisticated brain imaging equipment cannot reproduce pictures and stories from inside the mind of a dreamer, the dreamer's own account dictates how dreams are described in scientific journals. This has some obvious pitfalls, such as dreamers forgetting their dreams or finding them difficult to explain. We all know how it feels to try to grasp the pieces of a dream you have just awoken from – it seems as though the harder you try to recall the details, the faster they slip away. Another understandable problem is volunteers censoring their dreams for taste and decency. If a scientist asked you to relate a dream in which you knew you had murdered your mother or had sex with a horse, would you? Indeed, in Freud's day, and even until quite recently, you probably would have done better to withhold such dreams, as they would have been taken as indications that you were in need of psychoanalysis. Freud believed that every dream has a hidden meaning that had to be sought out, often one relating to his subjects' sexual fantasies. His influence on the field is impossible to ignore, even today, but we are finally moving away from his early understanding of dreams and towards one based on our developing knowledge about the brain and different states of consciousness (*see* QUESTION 5: WHAT IS CONSCIOUSNESS?).

By the 1970s, the US psychiatrist J. Allan Hobson had decided that Freud's theories had hampered scientific research on dreaming for too long and a more biological approach was needed. In April 1975, Hobson shared his views on dreaming with an audience at the University of Edinburgh, and two years later aired them in what became a controversial paper published in the *American Journal of Psychiatry*. Some considered this work a blatant attack on Freud. In essence, he and a co-worker, Robert McCarley, proposed that dreams were not psychological but physiological in origin and that their contents were more or less random. It suddenly seemed all too possible that dreams were nothing but the haphazard firing

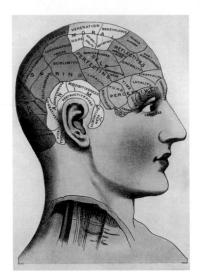

Until the early twentieth century, the time of Sigmund Freud, some scientists believed the brain was split into distinct areas that defined personality traits and could be mapped by a phrenologist who would feel for bumps on the skull. We are now able to map activity in different areas of the brain using imaging technologies, but the pseudoscience of phrenology has long since been discredited.

of active brain cells during sleep. Hobson and McCarley called the brain a "dream state generator" and suggested that our most bizarre dreams occur in the stages of sleep that we know as rapid eye movement (REM) sleep (*see* BIG DISCOVERY: REM SLEEP), so called because they are associated with fleeting motions of the eyes. We shift in and out of these periods of lighter sleep three or four times every night, as our minds fluctuate between states closer to waking consciousness and much deeper sleep. In the past, it was thought that dreams occurred only during REM sleep, but we now know that dreaming also occurs in deeper sleep – although coherent dream reports are much more difficult to extract from grumpy dreamers woken from deep slumber.

Taking an extreme stance against Freud's theories might lead us to the conclusion that dreams have no meaning or purpose whatsoever. However, research in the last couple of decades has disproved this idea. Thanks to developments in brain-imaging technology and the ingenuity of scientists in devising tests to probe the effects of sleep on mood, learning and memory, we are beginning to piece together a more complex picture of sleeping and dreaming. It now seems that dreaming may be a sort of "overnight psychotherapy", helping to keep our emotions in check, or even a virtual reality space for testing how real events might play out – a chance to simulate threats and our responses to them. Writing in 2010, Robert Stickgold and Erin Wamsley, sleep researchers at Harvard Medical School, US, called on scientists to "…abandon the entrenched view of dreaming as an unintelligible and mysterious phenomenon relying on entirely unknown brain processes and serving little or no function." Far from it, modern sleep research leads us to the far more intriguing conclusion that dreaming actually helps us in our waking lives and may be good for us (*see* INSIDE EXPERT: IS DREAMING GOOD FOR US?).

THE MEMORY MAZE

Many scientists now think that sleeping and dreaming are crucial to the brain's retaining and knitting together of memories, and some of the more striking recent findings in this area come from studies in animals. Although we cannot say for certain that animals experience dreams in the same way as humans, the patterns of activity in some animals' brains seem to suggest that they do – or at least that similar processes are going on in their brains, even if the animal is not aware of them. This is particularly useful in the case of rats because they are intelligent and can be trained to perform quite complex tasks. If we can observe how they process their experiences of these tasks in their sleep – in parts of their brains that deal with learning and memory – we may discover more about how our own waking experiences relate to the dream world.

"This one's useless - it keeps going, 'Oh, God, say it's only a dream.
Oh, God, say it's only a dream'."

← - -

A nightmare scenario! A *Punch* cartoon from the 1980s that reads: "This one's useless – it keeps going: 'Oh, God, say it's only a dream. Oh, God, say it's only a dream'."

In 2007, believing the answers to questions of dreaming could be found in the minds of rodents, neuroscientists at the Massachusetts Institute of Technology in Cambridge, US, reported on the results of sleep experiments involving rats navigating mazes. Matthew Wilson and Daoyun Ji had trained the laboratory-raised animals to run through these mazes in between short naps lasting one to two hours. Running through a maze has been a standard test presented to laboratory rats in learning and memory experiments for decades. Surprisingly, however, when Wilson and Ji looked closely at patterns of nerves firing in the rats' brains while they napped, some of them seemed to match almost perfectly with nerve-firing patterns recorded during the maze-running task. Wilson and Ji concluded that the rats were replaying their waking experiences in their sleep. It was as if their brains were using these "dreams" to try to memorize the maze or work out how to get around it better, and in fact, that is almost exactly what Wilson and Ji proposed was happening.

We can start to imagine the sleeping brain as a sort of curator of memories, flicking through the day's experiences, deciding which ones are important enough to keep and which collection they belong to in the "memory museum". During the curatorial process, two specific parts of the brain seem to be key: the hippocampus, an evolutionarily ancient part of the brain that collects new memories; and the cortex, a separate part responsible for storing memories long-term. Not only were these the same areas in the rats' brains where Wilson and Ji recorded the maze-running patterns, interplay between the two regions indicated that the memory-collecting and storing areas were communicating with each other, coordinating memory processing. Doing such studies in the laboratory has the advantage that you can control every element of the rats' lives, so you know how all of their experiences are formed and therefore

– if dreams are rooted in real-life experiences – where all their dreams are drawn from. On the other hand, when you wake up a rat it will never be able to tell you what it was dreaming about, whereas a person is often able to. So can this type of experiment be done in humans?

Actually, it can. Robert Stickgold's team of sleep scientists at Harvard have carried out a series of studies on people running through mazes in computer games. But instead of analysing their subjects' brain activities, the scientists simply observed them playing the games before and after a quick nap (and a chance to dream). Those who were allowed to sleep for an hour or so after learning how to play the game subsequently found their way around the computer mazes much faster than those who simply sat around and thought about it. And when the nappers were asked what they had dreamed about, some of them related scenarios involving mazes into which they had incorporated people they knew or other places they had visited in real life – in one case, a bat cave. This melding of new and old memories hints at a possible role for dreaming in placing new experiences in a wider context, and potentially helping us to plan for situations in our future that might resemble those in our past. In theory, the next time that particular dreamer gets lost in a bat cave, their brain should be able to retrieve the maze-navigating memories and offer some form of assistance. But the bat-cave dream also brings us back to one of the most enduring problems facing dream researchers: what if the bat

INSIDE EXPERT:
IS DREAMING GOOD FOR US?

"There's something about dreaming that seems to have a positive influence on our mood – it is like emotional homework that is important for our mental health. I've studied many, many people in the laboratory and knowing them before they sleep and what their issues are, and then collecting the dreams sequentially across the night, what I see is that they carry forward emotionally important material. While we're sleeping, we're tackling unresolved problems from our waking lives. Many studies have looked at what happens if you're prevented from dreaming, either by taking some medicine that suppresses that kind of sleep or because your dreams are interrupted in a laboratory. And what we've learned is that people who don't dream tend to be in a slightly worse mood when they wake up.

This function of emotional regulation is the only function of dreaming that is pretty well nailed down – it has had really consistent support from studies. But we assume dreaming has other functions. Even if you get up too early in the morning for a couple of weeks in a row, you will have had a reduction in your normal quota of dreaming and there will definitely be effects on your ability to learn, for example. So it appears that dreaming does have a real function and that we should honour it and get enough sleep so that we can complete our dreaming cycle."

Rosalind Cartwright,
Professor Emeritus in Neuroscience,
Rush University, Chicago, US

cave was not a bat cave at all? What if the dreamer just made it up, or changed it to remove embarrassing features? How can we ever really know why we dream unless we know what we dream?

After all this time, we are still recording dreams much in the same way that Maury did in the nineteenth century. Even in 2012, when reports emerged of a new technique that would allow scientists to read people's dreams, the reality was rather less sensational. Japanese researchers had found that by scanning the brains of volunteers in their sleep laboratory they could make a relatively reliable guess at what – from a list of 20 common dream objects – the volunteers were dreaming about. But although the scientists could often guess correctly whether or not a volunteer was dreaming about a car, they could not tell who, if anyone, was driving or where the car was headed. The dreamer would still have to fill in the story's details – if they could remember them. Dream research is hindered by the fact that for each one of us, a dream is an intensely personal experience and not – at least, not yet – one that we can simply record and replay as if it were a film. Like our waking thoughts, our dreams are ours and ours alone, but much, much more difficult to hold on to. We know the sleeping brain can act as curator, counsellor and creator. But the fact that dreams remain so out of reach, even to the dreamer, is just one of the reasons why science is still falling short of an altogether convincing explanation for why we dream.

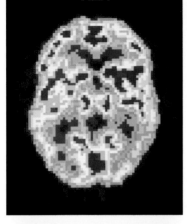

REM

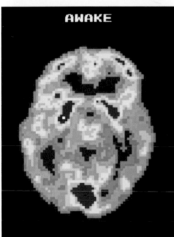

AWAKE

Scans of the brain during REM sleep and while awake. Warmer colours (red and yellow) indicate more active areas. The scans are generated through positron emission tomography (PET). The technique detects a radioactive "tracer" molecule, which is injected into the bloodstream and absorbed by the brain.

First detection of rapid eye movement (later termed REM) using electroencephalography (EEG), which records electrical activity in the brain. From Fig. 1 of the original Science paper by Eugene Aserinsky and Nathaniel Kleitman, published in 1953. - - →

BIG DISCOVERY:
REM SLEEP

Rapid eye movement (REM) sleep was first detected by US scientist Eugene Aserinsky in 1953, while he was still a graduate student at the University of Chicago, using his own son, eight-year-old Armond, as the subject. After connecting his son to a machine that recorded brain waves and eye movements, Aserinsky noticed that the traces recording eye movements were jerky, indicating wakefulness, even though his son appeared to be sound asleep. With his PhD supervisor, Nathaniel Kleitman, Aserinsky went on to co-author the paper in which REM sleep was first identified.

"…these physiological phenomena, and probably dreaming, are very likely all manifestations of a particular level of [brain] activity which is encountered normally during sleep. An eye movement period first appears about 3 hr after going to sleep, recurs 2 hr later, and then emerges at somewhat closer intervals a third or fourth time shortly prior to awakening. This method furnishes the means of determining the incidence and duration of periods of dreaming.".

From Aserinsky and Kleitman's paper "Regularly occurring periods of eye motility, and concomitant phenomena, during sleep", published in the journal *Science*, 4 September 1953.

RV

Resp.

RH

6:12 a.m.

Body

Motility

RF

RV = Vertical leads on right eye
RH = Horizontal "   "   "
RF = Right frontal (EEG)

Calibration: 200μv for RV and RH
50μv for RF

Paper speed: 10 seconds

Rapid Eye Movements During Sleep

7

WHY IS THERE STUFF?

I
n cavernous tunnels deep under the rolling hills of the French-Swiss countryside lies the "Big Bang machine". CERN's Large Hadron Collider (LHC) forms a ring 27 kilometres (about 17 miles) in circumference that spans the border between the two countries at a depth of more than 100 metres (about 330 feet). Using enormously powerful magnets, beams of subatomic particles are sent in opposite directions around the ring at close to the speed of light. When they collide, the particles smash together with such force that it helps scientists understand the high-energy conditions present in the tiny fractions of a second after the Big Bang. And it is here that physicists are searching for the answer to one of the greatest mysteries in physics: why we live in a universe of stuff.

In the instant that the Big Bang created our universe, there was only a sea of pure energy. And for that first millisecond, this energy was so great that it spawned numerous pairs of particles, each pair consisting of one particle of matter and one of antimatter. Matter is the "stuff" that makes up everything around us – including you. It's made of atoms which, in turn, have a core made of protons and neutrons and electrons whizzing around them. Protons have a positive charge and electrons have a negative charge, so the electrons stay in orbit around the core because opposites attract. Antimatter is the mirror image of normal matter. An atom of antimatter is the same as normal matter in every way, except that the charges are reversed: positively charged anti-electrons (known as positrons) orbit a negatively charged core of anti-protons and anti-neutrons. This isn't the stuff of science fiction – PET (positron emission tomography) scanning uses antimatter positrons to look inside the body and help diagnose diseases like cancer as well as produce images of how organs such as the brain and heart are functioning. A NASA telescope has detected antimatter being made in thunderstorms, and it is also created by the particle collisions in the LHC.

As the universe expanded and cooled after the Big Bang, the spawning of particle pairs ceased as the energy diminished. The amount of matter and antimatter became fixed. But that's only half the story. When a particle of matter encounters its antimatter equivalent (say, an electron and a positron), both turn back into pure energy in a process called annihilation. And since matter and antimatter are created in equal quantities, eventually all matter should annihilate with antimatter back into pure energy. However, then the universe would contain only energy – all matter would have cancelled out with all antimatter, leaving nothing from which to form galaxies, stars and planets. Yet the world around us is clearly filled with a lot of matter and not much antimatter. Your very existence is proof that the two must not have cancelled out perfectly. One possible explanation for this is that there is a reservoir of antimatter somewhere out there in the universe that hasn't yet annihilated with matter. However, surveys of the universe show no evidence of such hidden antimatter out to a distance of 10 billion light years (about 95 billion trillion kilometres, or 58 billion trillion miles). An alternative explanation is that matter and antimatter aren't perfectly symmetrical after all, and so they didn't quite cancel out after the Big Bang. Unable to recreate the exact conditions that created our universe, physicists instead use particle accelerators like the LHC to reproduce the conditions as closely as possible in order to test these ideas. These experiments can produce an additional suite of subatomic particles that we don't encounter day-to-day because they appear only in high-energy environments like the Big Bang. Physicists can then look for evidence of the symmetry between matter and antimatter breaking down. Such evidence has already been found in particles called kaons.

First discovered in 1947 at the University of Manchester, UK, kaons are unstable and so decay quickly into other particles. This can happen in many different ways, one of which creates a positron, while another creates an electron. If matter and antimatter really are symmetrical,

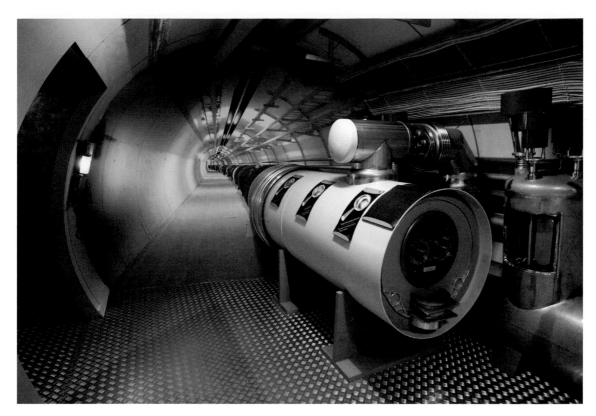

A model of the Large Hadron Collider (LHC) at CERN, near Geneva, Switzerland. Beams of sub-atomic particles are sent in opposite directions around a 27-kilometre (16-mile) long underground track and smashed together at close to the speed of light.

then even over millions of decays, physicists should expect to find equal numbers of positrons and electrons created. However, experiments have shown that positrons are created 50.17 per cent of the time, compared to 49.83 per cent for electrons. Physicists call this breaking of the matter-antimatter symmetry "CP violation". The first experimental proof of it was provided by scientists working at the Brookhaven National Laboratory, Long Island, New York in 1964. For their groundbreaking discovery, US physicists James Cronin and Val Fitch were awarded the 1980 Nobel Prize in physics. It is this process of CP violation that could hold the key to answering the mystery of why we live in a universe of stuff. Yet, a discrepancy in kaons alone isn't enough to account for the matter-antimatter imbalance in the universe. Enticed by the findings of Cronin and Fitch, physicists working at CERN hunted for more instances of CP violation.

SUPERSYMMETRY

Around the LHC's circumference sit four cathedral-sized spaces, one of which houses the LHCb experiment. At 21 metres (about 69 feet)

$$M_{\pi\pi} = (498 \pm 10 \, mev)$$

# of Events

50 events

10 events

0

500

US physicists James Cronin (left) and Val Fitch (above) shared the 1980 Nobel Prize in physics for their discovery of CP violation – a breaking of the symmetry between normal matter and its mirror counterpart antimatter.

long, 13 metres (about 43 feet) wide, 10 metres (about 33 feet) high and weighing 5,600 tonnes, LHCb is a monster machine, but a very sensitive one – like its three fellow detectors, it is capable of sifting through the remnants of the LHC's epic collisions. Some of those collisions create particles known as D mesons, as well as anti-D mesons – matter and antimatter. Like kaons, both are unstable and rapidly decay into lighter particles that LHCb's finely tuned sensors can detect. And by extrapolating backward, physicists can work out whether these particles initially came from a D meson or an anti-D meson – whether they were born from matter or antimatter. Based on the near-symmetrical nature of matter and antimatter, physicists expected the D mesons to break down in the same way as the anti-D mesons. According to the standard model of particle physics (the most widely accepted theory concerning forces, particles, and their interactions at the subatomic scale), the difference between the two – the amount of CP violation – would be only 0.1 per cent. However, results published in November 2011 showed a discrepancy of 0.8 per cent – eight times what was expected. It seems there is more CP violation going on in the LHCb than physicists had anticipated.

What could explain this surprising result? One possibility is a theory called supersymmetry, which has been around since the 1970s. It predicts that every particle has a bigger, heavier superparticle, called a "sparticle" (an electron would have a sparticle known as the selectron, for example). It is possible that sparticles are disrupting the decay of normal particles like the D mesons and anti-D mesons in the LHCb. Sparticles are so heavy that they require huge amounts of energy to produce, but, according to the laws of physics, such huge amounts of energy can be "borrowed" from nature as long as the debt is swiftly repaid. So it's possible that pairs of sparticles and anti-sparticles are popping into brief existence within the LHC, before annihilating and disappearing again incredibly quickly. Physicists have modelled the potential effect of these short-lived sparticles, and the calculations show that they could produce a discrepancy as much as 1 per cent, which is close to the 0.8 per cent actually measured.

If this really is the first evidence of supersymmetry, it would change the landscape of physics. It would also explain many of its biggest mysteries. For one, the lightest of the superparticles could be responsible for the dark matter that seems to make up 27 per cent of our universe (*see* QUESTION 1: WHAT IS THE UNIVERSE MADE OF?). Supersymmetry also plays a role in the physics of higher dimensions, which might one day tell us what happens at the bottom of a black hole (*see* QUESTION 17: WHAT'S AT THE BOTTOM OF A BLACK HOLE?). However, scientists are careful not to jump too hastily to conclusions, so they went back to double-check how they got that original prediction of 0.1 per cent. They found that they had underestimated the contribution of certain factors, so the standard model could explain a discrepancy of 0.8 per cent after all. So, for now, there is no conclusive proof that supersymmetry exists. Experiments at the LHC will continue to look for evidence of supersymmetry, and the LHCb team will continue to look for more instances of CP violation. And yet many physicists believe that the CP violation observed at CERN still won't completely explain why we live in a universe of stuff. In the search for another piece in the puzzle, attention is turning to an even tinier particle: the neutrino.

GHOST PARTICLES

Neutrinos are ghostly particles made in high-energy environments like the centre of the Sun. In fact, around 65 billion solar neutrinos pass through every square centimetre (about 420 billion per square inch) of your body every second. That you don't notice this flood of subatomic particles streaming through you shows just how little they like to interact with normal matter. Yet despite this lack of interaction, a tiny fraction of the neutrinos passing through the Earth can be detected using the right equipment. And it was through investigating these particles that physicists in the 1980s noticed something was missing: the number of neutrinos arriving from the Sun seemed to be about a third of the number expected.

This mystery remained unsolved until physicists realized they were wrong in initially assuming neutrinos had no mass. By the late 1990s, they'd found that the neutrino did have a mass – a really tiny one, at least a million times less than that of the electron. And it turns out that this small amount of mass allows a neutrino to shape-shift ("oscillate") into three different varieties. The mystery of the "missing" neutrinos was solved: on their journey from the Sun to Earth they had changed into the other two varieties, which the equipment could not detect.

It is these oscillations that might hold the key to unlocking the mystery of the matter-antimatter imbalance. CP violation – the breaking of the symmetry between matter and antimatter – has never been observed in neutrinos and anti-neutrinos. Their reluctance to interact with normal matter means that they are rarely detected and it will take time to gather enough data to see whether CP violation does, in fact, occur. However, it is possible that neutrinos and anti-neutrinos do differ in the way they shift between their three varieties. Experiments like the T2K particle-

A giant magnet that forms part of CERN's LHCb detector. Results published in November 2011 revealed that more CP violation was occurring during particle collisions in the LHCb than had been anticipated, leading to speculation about the theory of supersymmetry.

accelerator experiment in Japan, will, in the near future, look for these differences as the particles swap variety (*see* INSIDE EXPERT: HOW CAN WE LOOK FOR CP VIOLATION IN NEUTRINOS?). We currently have little idea how much CP violation to expect in neutrinos – it could be none, a little or a lot.

The neutrino experiments and those at the LHC are allowing physicists to attack the same question from several angles and hopefully discover an answer to the elusive question of why our universe seems to have a preference for matter over its antimatter counterpart. It is one of the major mysteries of modern science and solving it may finally explain just why there is stuff in the universe at all.

INSIDE EXPERT:
HOW CAN WE LOOK FOR CP VIOLATION IN NEUTRINOS?

"Neutrinos behave strangely. Well, strangely to you and me but quite normally for quantum particles. As a particle travels from birth to death, you can only say what it might be like. You can only give a probability that it exists at a certain point in space and time and has certain properties – that's it. Only when measured are its location and properties pinned down. Usually, the time between interactions is so small that the chance of a particle changing any of its properties is tiny. But neutrinos have a lot more time to change because they don't like to interact with matter – and change they do.

Neutrinos are born in one of three types: electron neutrino, muon neutrino or tau neutrino. Each corresponds to the three charged particles they are associated with. When a neutrino interacts with matter, it produces its corresponding charged particle. So if a neutrino has changed 'flavour', experiments will see the production of a charged particle that's different from the one corresponding to its original state. This whole process of change is called neutrino oscillation. We have measured the way neutrinos oscillate with many experiments using different sources. This has been achieved by looking at the probabilities that one type of neutrino chooses to change into another. Now

it's time to see if neutrinos act in the same way as their antimatter versions: anti-neutrinos.

Nature seems to prefer matter to antimatter at a very small level, and without this bias none of the visible universe would exist. By carefully observing the oscillation of neutrinos and anti-neutrinos, we hope they display some difference in behaviour. Future neutrino experiments will fire beams of neutrinos, and then anti-neutrinos, hundreds of kilometres [hundreds of miles] through the Earth to massive detectors. When comparing results, both between neutrinos and anti-neutrinos as well as across other experiments, we will be able to tell us if they behave in different ways – whether CP violation occurs in neutrinos and anti-neutrinos.

If differences are seen, this could be evidence for nature's matter bias. It will take some time, though, to see enough ghostly neutrinos and anti-neutrinos to really understand if they can explain the unlikely presence of stuff in the universe we live in."

Dr Ben Still,
neutrino physicist,
Queen Mary, University of London, UK

8

ARE THERE OTHER UNIVERSES?

Y ou shouldn't exist. The fact that you do flies in the face of all probability. In fact, in some ways it seems our universe has been made just so that you can exist. So how did you arrive here, now holding this book in your hand? You were probably born approximately nine months after the amorous activities of two *Homo sapiens*, a species that has evolved to be the most dominant on planet Earth. Our planet formed around 4.6 billion years ago as a by-product of the formation of the Sun from a collapsing cloud of interstellar gas. In turn, that gas cloud formed from the amalgamation of stellar material ejected into space by the explosive death of several big stars. Contained within that cloud were some of the heavy elements that are now found in you: the calcium in your bones and the iron in your blood.

But it could have been very different. Tinker with the settings of nature just a little and things don't work out so well. If the strength of certain physical forces were altered even slightly, you simply wouldn't exist. Make gravity a few per cent weaker and it wouldn't create sufficient pressure in the hearts of stars to fuse the lighter chemical elements into heavier ones, so there would be no calcium or iron in our universe. If the force that binds the centre of atoms together was a little stronger, then the Sun would consume all of its fuel in a few seconds, nowhere near enough time for life to evolve. Increase the force that governs radioactivity and stars wouldn't explode at all, so the ingredients that make up your body could not have been flung across space to mix and collapse, putting off the chain of events that led you to these pages. The improbable, life-friendly balance of nature might suggest that our universe has been fine-tuned in order to allow us to exist. Some will argue that this points to a Creator who intentionally designed things this way, but there is another explanation: that our universe isn't the only one.

Imagine that there are countless other universes out there – a multiverse – with the strengths of the physical forces varying slightly in each. In some, the forces are only fractionally different from ours; in others, vastly so. Every possible combination of settings is played out somewhere across the multiverse. In some of the universes, gravity really will be weaker and calcium and iron will never form. In others, stars will live and die in the blink of an eye. Amid this infinity of universes, which one would you expect to find yourself in? The answer, of course, is the one where the settings allow your existence. The reason our universe seems to be so fine-tuned to life is that, if it wasn't, we wouldn't be here to notice. This notion may seem like a philosophical "get-out-of-jail-free card", a clever trick to avoid having to answer the real question. But the remarkable thing is that other universes aren't just science fiction – some of physics' most widely accepted theories actually predict their existence. The first of these theories requires going right back in time to the birth of our universe: the Big Bang.

The Big Bang theory developed in the wake of US astronomer Edwin Hubble's discovery in 1929 that the universe is expanding. If the universe is getting bigger now, it must have been smaller in the past – it had a beginning. In the decades after Hubble's discovery, astronomers found evidence to suggest the Big Bang was an accurate description of this origin. However, by the 1970s, several problems with the theory had emerged. For one, it struggled to account for the apparent flatness of the universe. Flat is one of three possible alternatives for the universe's overall shape; the others are closed (curled up like a sphere) and open (shaped like a saddle). Which shape the universe actually is depends on how much stuff it contains. If the universe has a lot of stuff, it is closed; if it has little, it is open. Only if it has just the right amount of stuff – the so-called critical density – will it be flat. Measurements suggested that the amount of stuff in our universe is, indeed, very close to this critical density, so once again astronomers were faced with the uncomfortable coincidence of fine-tuning.

By the early 1980s, US physicist Alan Guth and other scientists provided a possible solution to the "flatness problem". They proposed that, very shortly after the initial Big Bang, our infant universe underwent a period of superfast inflation – it expanded by a factor of 10^{78} (1 followed by 78 zeroes) in a tiny fraction of a second. Inflation can account for our universe's flatness because any original curvature it had would have been

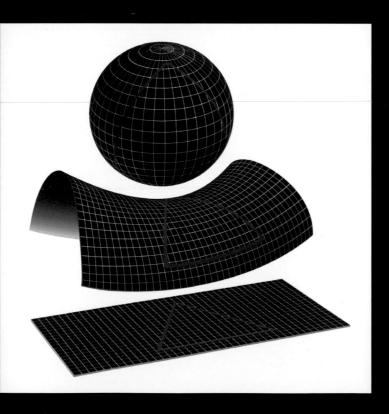

← - -

The three possible shapes for the universe: closed (top), open (middle) and flat (bottom). Only a precise amount of stuff in the universe – the critical density – would make it flat, and yet that's what is appears to be. This flatness problem led, in part, to the theory of inflation.

smoothed out as space rapidly stretched. It also raises the possibility of the existence of other universes.

POCKET UNIVERSES

According to inflationary theory, the culprit responsible for the cosmic growth spurt in the early life of the universe is the inflationary field. As a subject, physics is littered with similar fields: variations in the electromagnetic field give rise to light; the gravitational field is responsible for the force of gravity; even the famous Higgs boson is a manifestation of the Higgs field. Inflation is seen as a manifestation of a field called the inflaton. It is energy locked up in this field that inflationists believe causes expansion to accelerate rapidly. However, the theory also suggests that the strength of the inflaton field varies. In areas where the field is strong, inflation continues; where the field is weaker, inflation stops. So overall, the universe continues to inflate but "pockets" are left behind where inflation has ceased. These pockets are, for all intents and purposes, separate, distinct, isolated universes. What's more, the laws of physics will differ between them depending on the exact way each pocket formed. If this theory is true, we must live in such a pocket universe – a place where superfast inflation has stopped – because astronomers simply don't observe any such superfast inflation in our universe today, although it is still expanding at a considerably slower pace. If, as believed, our universe is just one of an infinite number of these pocket universes, everything that can happen does happen somewhere in the multiverse. In fact, it happens an infinite number of times. Your existence no longer becomes highly improbable, it becomes inevitable. It comes as no surprise, then, that we reside in a pocket universe where the strengths of the fundamental forces became fixed in such a way that life could one day develop – you couldn't find yourself in a universe where they hadn't.

It may sound like physicists are trying to use a sledgehammer to crack a nut. That they are invoking inflation, and its requirement for an infinite number of pocket universes, just to paper over a few cracks in the Big Bang theory (like the flatness problem). However, inflation, like any scientific theory, lives or dies on whether it can make testable predictions about our own universe. It makes such predictions about a type of radiation known as the cosmic microwave background (CMB) – the afterglow of the Big Bang – and precise measurements of the CMB made by space telescopes are in near-perfect agreement with the predictions of inflationary theory. This currently makes inflation the best explanation for the way our universe looks, and multiple universes, perhaps the reason for your very existence, come packaged up with it. However, inflation isn't the only well-tested theory in

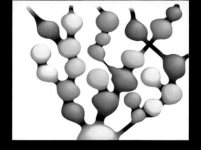

The inflating universe. Overall, inflation continues (from bottom to top) but pockets get left behind where local inflation has stopped. If inflation theory is true, we must live in one of these pockets because astronomers don't observe such super-fast inflation in our universe today.

Artist's impression of some of the pocket or bubble universes thought to make up the multiverse. Each bubble is totally isolated and the laws of physics differ between them. It is only possible to find yourself in a bubble where those laws allow life to exist.

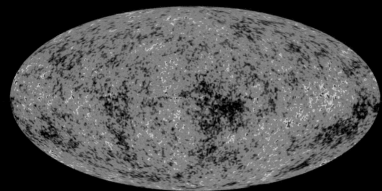

A map of the cosmic microwave background (CMB) created from nine years of observations by the Wilkinson Microwave Anisotropy Probe (WMAP). The map matches specific predictions made by inflation, lending weight to the theory.

← – –

physics that leads to the possibility of other universes. Quantum physics has something to say on the matter too, and it all starts with an imaginary cat.

A CAT IN A BOX

In 1935, Austrian physicist Erwin Schrödinger was trying to better understand quantum physics. He imagined the following scenario: a cat is trapped in a box, in the corner of which is a vial of poison and a radioactive material. If an atom from the radioactive material decays, the vial breaks and the cat is killed. If there is no decay, the cat lives. Whether or not an atom decays is random and is governed by the rules of quantum physics, which differ considerably from those of our everyday experience. For example, until it is measured, an atom cannot be said to be in one particular place. In fact, it simultaneously exists in several places at once – known in the language of physicists as being in a superposition of states. This is a strange but very well-tested idea. However, when we look at an atom, we see it in one specific location. The most widely accepted explanation for this is that it's the very act of observation itself that forces the atom to "choose" to be in that location. This explanation is known as the Copenhagen interpretation, named after the city where the Danish physicist Niels Bohr formulated the idea. Schrödinger came up with the imaginary cat experiment to push Bohr's idea to its limit, and what emerged has implications for the possibility of other universes.

According to the original Copenhagen interpretation, an atom in the radioactive source remains in a superposition of states – a combination of decayed and not decayed – until someone opens the box to make an observation. Only then does it unambiguously become one or the other. However, the logical implication is that, until someone opens the box, the cat is both alive and dead. It, too, is in a superposition of states. It is hard enough to accept atoms being in several states at once. It is downright bizarre to imagine a zombie cat that exists in a hybrid state of dead and undead. This counter-intuitive idea has been debated many times since Schrödinger first imagined it. Subscribers to the Copenhagen interpretation see no problem, simply defining an observer as anything or anyone who is looking. The very fact we've installed a detector inside the box to look for radioactive decay is enough to force the atoms to "choose". But critics of the Copenhagen interpretation argue that it says nothing about why an observation must force a choice. What if it doesn't? What if both outcomes occur? One man thought just that.

One night in 1954, 23-year-old physicist Hugh Everett III was drinking sherry with colleagues at Princeton University and debating the implications of the Copenhagen interpretation. By then, this interpretation was widely accepted because its equations made remarkably accurate predictions about what physicists see when looking at the behaviour of light and atoms. But those same equations

INSIDE EXPERT:
WHY SHOULD WE BELIEVE IN MANY WORLDS?

"Many Worlds theory has been around since 1957, but in the last 25 years it's become popular with large numbers of physicists. Over that period, the Copenhagen interpretation – for 50 years the overwhelmingly dominant interpretation of quantum theory – has lost its dominant status.

Why the change? While Bohr and the other Copenhagen founders really did think hard about the theory, most physicists in the post-war period were more concerned with using quantum theory to do calculations. It wasn't really necessary to understand the meaning of quantum theory, and so it was convenient to pass lightly over the conceptual puzzles.

That changed in the 1980s for several reasons. Firstly, scientists started to directly see the weirdness that quantum theory predicts and so it couldn't be ignored. Also a new theory – called decoherence – changed the way quantum theory was looked at. The Copenhagen interpretation tries to reconcile the differences in behaviour between objects in our everyday world and those in the quantum world by saying that observation is important. However, decoherence theory provides a natural way in which the behaviour of our macroscopic world emerges out of the quantum world without the need for observation.

It says that, as far as quantum particles are concerned, there's just one, tangled-up, very weird reality. But, at larger scales, that reality untangles and separates out into layers that basically don't interact. Those layers are the "worlds" of the Many Worlds theory. Decoherence theory tells us how these layers are created out of the single, mixed-up reality at the quantum level. So the Many Worlds theory doesn't add some extra many-worlds physics to existing quantum theory. It is actually telling us that quantum theory, taken literally, already is a many worlds theory!"

Dr David Wallace,
Philosopher of Physics,
University of Oxford, UK

also tell us that two versions of the cat simultaneously reside inside Schrödinger's box. Bohr tried to reconcile the success of the equations with the fact that we would only ever observe one cat, by making the act of observation important. But, that night in 1954, by questioning the need for observation, Everett hit on a new interpretation. Later, he was able to show that observation was no longer needed for the equations to produce the reality we actually experience. However, Everett's interpretation required something radical: multiple universes.

TWO CATS IN TWO BOXES

According to Everett's idea – now called the Many Worlds interpretation – when faced with a quantum "choice", the universe splits. In the case of Schrödinger's cat, the whole universe splinters into two versions, one with a live cat, another with a dead one. As we can perceive only one universe, we only experience one of these branches as "reality". However, if you see a dead cat, there is an equally real version of you in another universe who discovers the opposite. This has implications for our original fine-tuning problem. Due to the sheer number of quantum events occurring every second across the entire cosmos, the universe would be constantly splitting off an infinite number of isolated branches. Once again, everything that can happen does happen. Your existence, and your reading of these pages, is once more inevitable.

If you are struggling to accept this idea, then you are not alone – when Everett presented his ideas in 1957, he was encouraged to publish a severely abridged version. Even then, some physicists labelled his idea of constantly fracturing universes "repugnant". His work all but discredited, Everett quit physics and went to work for the US military at the height of the Cold War. A chain-smoking alcoholic, he died of a heart attack in 1982, aged just 51.

It was only after Everett's death that more than a handful of physicists started to take his ideas seriously. Today, the Many Worlds interpretation is almost on a par with the Copenhagen interpretation as the explanation of choice among modern physicists – the best way to reconcile scientific theory with our actual perception of reality (*see* INSIDE EXPERT: WHY SHOULD WE BELIEVE IN MANY WORLDS?). Distinguishing between the two interpretations is difficult because they mostly make the same predictions, but there have been suggestions for how to test the Many Worlds interpretation. The most extreme of these – a variation of Schrödinger's cat experiment known as the quantum suicide experiment – has been most discussed by Swedish physicist Max Tegmark.

In this thought experiment, a scientist sits with a gun pointing at their head. The weapon is wired to a detector that measures protons. A proton, a subatomic particle, has a property called spin, which is "up"

$$i\hbar \frac{\partial \psi}{\partial t} = -\frac{\hbar^2}{2m} \nabla^2 \psi + V\psi$$

The Schrödinger equation of quantum physics. The equation makes remarkably accurate predictions about how atoms and light behave in experiments. However, arguments still rage about how to interpret it in order to produce our experience of reality.

exactly 50 per cent of the time, and "down" the other 50 per cent. If the detector measures "up", the gun fires and kills the scientist. If it measures "down", the gun merely gives out an audible click. Under the Copenhagen interpretation, after one measurement you have 50 per cent chance of being dead; after two clicks, a 75 per cent chance; three clicks, an 87 per cent chance and so on. Why would anyone attempt such a crazy experiment? Well, some researchers believe the Many Worlds interpretation says you will never die. After the first proton measurement, the universe would splinter into two, one in which you die and one where you live. Unable to perceive a universe in which you are dead, you must experience the one in which you survived. A proton measurement in that universe would subsequently divide the universe once more and you could only continue to perceive the universe in which you are not shot. There will always be a branch of the Many Worlds tree where you have, against all odds, survived hundreds of measurements and that is the only universe where you can logically find yourself. So according to the Many Worlds interpretation, if your life hinges on quantum measurements then your existence is not just inevitable, you are also immortal!

The extreme nature of this imaginary experiment just goes to show the difficulty in separating the Many Worlds and Copenhagen interpretations. Whether one will win out over the other remains to be seen. But if the Many Worlds interpretation is correct, along with the inflationary multiverse it points to there being an infinite number of yous in an infinite number of universes. There will be some where you were never born and others where you live to be 200. In others, you'll be a Nobel Prize winning scientist who single-handedly answers every question in this book; in yet others, they will forever remain unanswered. Such could be the nature of reality.

9

WHERE DO WE PUT ALL THE CARBON?

Our children's children will grow up in a world we barely recognize. Not just because of the relentless progress of technology and the fickle nature of fashion, but because the planet itself is changing. We are facing a slow crisis: the environment is gradually metamorphosing and the natural world we know today is disappearing. The greenhouse gases that our power stations pump into the atmosphere – the biggest culprit being carbon dioxide – are trapping energy from the Sun and changing our climate. Unless we stop burning fossil fuels and make our energy some other way, or find somewhere to safely lock away all the carbon, the problems that come with climate change will only get worse. Future generations will live with more frequent storms, floods and droughts, and, as ancient ice melts, rising seas will force coastal communities to retreat more rapidly inland (*see* BIG DISCOVERY: WHAT HAPPENS WHEN AN ICE SHELF WARMS UP). Even if we can cure our addiction to fossil fuels, greenhouse gases like carbon dioxide could hang around for hundreds or thousands of years, warming our world for future generations. Accepting that this is happening is one thing; working out what to do about it is quite another.

The pace of climate change, which occurs over the course of entire lifetimes, makes it hard to come to terms with. Perhaps we can appreciate it better from the perspective of someone, or something, that has spent far longer on this planet than we have. On the slopes of the White Mountains in California lives a single, sturdy bristlecone pine tree that has survived for nearly 5,000 years. Named after the long-lived biblical figure, it is known as "Methuselah". When the seed of this ancient tree germinated, carbon was recognized only in the form of charcoal or soot. In its lifetime the Roman Empire has risen and fallen. Only in the last couple of hundred years – a mere blip in the twilight of Methuselah's life – has industrialization produced the carbon that is choking our atmosphere and changing our weather patterns. Methuselah has barely shed one coat of needles since talk of global warming began in the 1970s. So while from our point of view climate change may seem to be creeping up on us so slowly that it is barely happening at all, the reality is that the face of the planet is being altered in the briefest moment of our history.

Carbon exists in our atmosphere as carbon dioxide; it is supposed to be there and it is essential to plants, like the Methuselah tree, that use it to make their energy. The ceiling for safe levels is thought to be around a trillion tonnes. If this estimate is correct, when more than a trillion tonnes are emitted, the accompanying rise in temperature will exceed 2 degrees Celsius (3.6 degrees Fahrenheit) above pre-industrial levels. Scientists around the world have developed complex models to try to understand what this will mean for our future; they predict more extreme weather events, health problems related to heat and air pollution, changes in the way disease spreads, to name but a few of the uncomfortable consequences (*see* INSIDE EXPERT: WHAT HAPPENS IF IT GETS TOO HOT?).

And it seems likely that the burden of climate change will fall most heavily on those who are least able to adapt – the poor.

By burning fossil fuels, we have already unleashed around half a trillion tonnes of carbon and now need to reverse the upward trend in emissions, making cuts of a few per cent each year, to avoid releasing the other half. We are like a company that has spent half of its budget for the year on staff bonuses and is trying desperately to avoid collapse. Even as richer countries try to curb their carbon expenditure, newly industrialized countries – undergoing the same processes of economic and technological development that led to climate change in the first place – are starting to make deeper holes in the budget. Many scientists now think we need to start taking carbon out of the atmosphere and locking it away in the land, or under the sea, to avoid the worst effects of climate change – even if cleaner energies are just around the corner (*see* QUESTION 10: HOW DO WE GET MORE ENERGY FROM THE SUN?). Simply planting more trees will not solve the problem. Methuselah and other land plants may ease some of the strain by absorbing carbon dioxide from the air, but they have so far managed to mop up less than a third of what we have released in the last 150 years. The oceans have made a similar contribution. Instead, we need clever storage solutions, and this is where the science gets controversial.

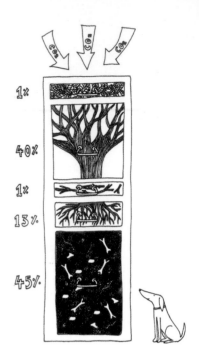

1%

40%

1%

13%

45%

Where is the carbon in a tree stored? It depends on the tree, but figures for a Great Lakes ecosystem suggest much of it is stored below the ground. From top drawer to bottom: leaves, branches, twigs, roots, soil.

DEEP STORAGE

The perfect place to put the carbon is essentially back where it came from. Empty oil wells and gas fields are prime sites for underground or underwater carbon storage. The idea is to capture the carbon dioxide liberated from fossil fuels and plumb it straight into storage sites, reducing the amount of carbon that power stations belch out into the atmosphere. The technology for this already exists. In the middle of the North Sea, the Sleipner West oil and gas field operates a carbon capture and storage (CCS) facility where around a million tonnes of carbon dioxide are piped under the seabed every year. Operational since 1996, the project was the first of its kind and remains one of the deepest at more than 800 metres (2,625 feet) below the seafloor. But CCS is useful only while fossil fuel sources remain, and this poses something of a dilemma: should we be spending our time and money trying to squeeze every last drop of energy from fuels that are rapidly running out, when we could be investing in greener alternatives instead? For the five years to 2012, the solar power industry was growing around ten times faster than the carbon capture and storage industry, which suggests that some investors have already placed their bets.

An even more controversial option is sprinkling the sea with iron to encourage the growth of carbon-consuming algae. Like plants, algae called phytoplankton rely on carbon dioxide from the air to generate their energy through photosynthesis. They naturally form large surface blooms – so large

Carbon dioxide (CO²) can be captured directly from the air, or in the gases produced from burning fossil fuels, or by heating coal to release carbon monoxide – which produces CO² in a reaction with water. It is liquefied before being transported to storage sites.

that they are sometimes visible from space. But these blooms can also be triggered artificially by adding iron to patches of ocean; the iron acts like fertilizer, leading to a population explosion of phytoplankton. In theory, when the algae eventually die, they sink, taking their carbon with them to the bottom of the ocean. However, we know relatively little about the workings of marine ecosystems, and in particular about the workings of the deep sea, so it would be hasty to assume that large-scale fertilization of the oceans would have the desired effect or that it would have no unwanted side effects. Indeed, scientists wishing to carry out large-scale iron fertilization experiments in the open ocean must navigate a minefield of legislation designed to prevent marine pollution. So far, only a handful of small and medium-sized experiments have sought to probe the potential of iron fertilization and it remains uncertain how long carbon sunk to the seabed would stay there.

Other options for locking away carbon in the deep sea are flawed for similar reasons. By taking advantage of its heavier-than-water state in deep water, carbon dioxide could be pumped directly to the bottom of the ocean without having to be contained under rocks or in decaying matter. At a depth of around 3 kilometres (about 2 miles), the gas mingles with seawater to form a substance called clathrate (the same stuff is thought to be trapped in the surface layers of Mars). But the effects on deep-sea dwellers are unknown and leakage could return as much as a quarter of the clathrate carbon to the surface within a century. Between 2004 and 2006, a team of US scientists carried out a series of experiments to try to predict the possible impact of deep-sea carbon stores on ancient, mud-dwelling life forms called foraminiferans, which, although small, are found all over the seafloor and provide sustenance for species higher up the food chain. The scientists injected carbon dioxide directly into mud on the seafloor off the Californian coast and returned weeks later to find out the effect on the foraminiferans living there. They found that while some species remained unscathed, the shelled species had been wiped out by clathrates. It turns out that the clathrates had created an acidic environment, affecting chemicals that are key components in shell-building and making them unavailable to the foraminiferans.

The acid problem is not limited to deep-sea species. As the oceans absorb more of the carbon dioxide from our atmosphere, they are becoming increasingly acidic. Coral reefs and squid are extremely sensitive to changes in acidity and will suffer as a result, but our knowledge of the deep ocean is so limited (*see* QUESTION 16: WHAT'S AT THE BOTTOM OF THE OCEAN?) that the full extent of the problem is still emerging. One surprising consequence of the ocean absorbing more carbon dioxide is that it will become noisier. As seawater becomes more acidic from the increased carbon dioxide, its ability to absorb sound is reduced and noises travel further, which may affect animals like whales that rely on sound to hunt and navigate and fishermen who use sonar to locate fish.

HIDE AND SEEK

Faced with a shortage of quick-fix solutions, it is easy to become disengaged from the climate change debate and hope instead that nature finds some way to right our wrongdoing. Nature is resilient and has some neat, built-in safety mechanisms – as the ocean becomes more acidic, for instance, its capacity for taking up carbon decreases – but it is far from immune. In many ways, the problem is self-perpetuating; the effects of climate change are unlocking natural storage sites that have been keeping carbon safe since before the time of the Methuselah tree. Take, as one example, peat bogs. In 2011, scientists at Bangor University in Wales calculated that peat bogs in Britain alone could hold more than three billion tonnes of carbon, as well as other important greenhouse gases like methane. Worldwide, peatlands may contain twice as much carbon as all the trees in all the forests. As a result of climate change, however, it is predicted that peat bogs will dry out, releasing their carbon contents – an effect that will worsen as climate change tightens its grip. Seagrass is another example. Ancient seagrass meadows in the Mediterranean Sea are home to plants that may have lived as long as 200,000 years, making Methuselah look positively infantile. But it is what is they are hiding that is important. For every hectare (about 2.5 acres) of seagrass, it is thought there may be millions of tonnes of carbon trapped in the underlying

- - →

The Larsen B ice shelf on 7 March 2002, following its collapse. The blue patch is a mixture of slush and icebergs.

BIG DISCOVERY:
WHAT HAPPENS WHEN AN ICE SHELF WARMS UP

In February 2002, scientists keeping an eye on the Antarctic via satellite imagery spotted something extraordinary – and profoundly sad. Larsen B, an ice shelf the size of a small country, was disappearing before their eyes. The world looked on helplessly as 3,250 square kilometres (1,255 square miles) of ice that had been frozen for thousands of years cracked and disintegrated in a matter of days. The ice shelf's demise was mourned by the rock band British Sea Power in their song "Oh, Larsen B": "Five hundred billion tonnes, Of the purest pack ice and snow, Oh Larsen B, Oh, won't you fall on me?" It seemed clear that rising temperatures had contributed to the collapse, and scientific support was lent to this idea in a study published in 2005:

"...the recent collapse of the Larsen B ice shelf is unprecedented... [We] suggest that the recent prolonged period of warming in the Antarctic Peninsula region, in combination with the long-term thinning, has led to collapse of the ice shelf."

From "Stability of the Larsen B ice shelf on the Antarctic Peninsula during the Holocene epoch", published in the journal *Science*, 4 August 2005.

Floating ice shelves like Larsen B do not directly affect sea levels because they already displace water, but their loss exposes glaciers and ice sheets that could in turn drive sea level rise. Under the influence of climate change, it is possible that the West Antarctic Ice Sheet covering West Antarctica will melt, freeing enough water to cause a global sea level rise of 3 to 6 metres (10 to 20 feet) and leaving low-lying islands like the Maldives, and many coastal cities, under the waves.

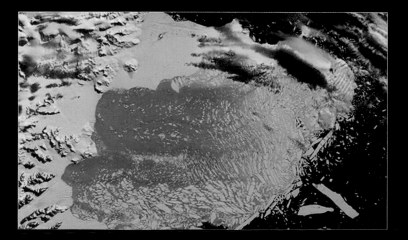

seafloor. Even if seagrasses cover only a tiny fraction of the seafloor, they may still be as important for storing carbon as our forests – and both are disappearing. Scientists predict that Mediterranean seagrass beds will be almost non-existent by around the middle of this century, even if we manage to stabilize our greenhouse gas emissions by that time.

The carbon conundrum is fast becoming a game of hide-and-seek. Even as we stash carbon away under the North Sea, we open up century-old stores of it by draining peat bogs in Indonesia. The trouble with global warming is that it is just that – global – and the actions of one community, one company or one country have an impact on us all. Positive steps taken to tackle climate change can very easily be undermined. We know that we share a global responsibility to look after our atmosphere, and our future, but galvanizing local efforts remains difficult in the face of such overwhelming challenges.

When Methuselah's last needle falls, where will we be? Our forecasts of future climate scenarios are only as good as the numbers we feed into them – numbers that predict where the carbon balance of our future planet will lie. Many scientists now believe that exceeding the two degrees Celsius (3.6 degrees Fahrenheit) temperature limit is almost inevitable. Meanwhile, the ancient tree serves as a stark reminder of the uncertain ends to which climate change may lead us. Though bristlecone pines like Methuselah thrived in the age of woolly mammoths (the Pleistocene Epoch), that was a cooler, wetter time. Bristlecone pines have continued to occupy the slopes of the Great Basin in North America but have slowly found themselves confined to loftier heights, and bristlecone forests that

As well as storing more carbon than forests, peat bogs are important habitats for plants. Labrador tea, bog rosemary, wild blueberry and sphagnum moss thrive in the peat bogs of the Mingan Archipelago in Canada.

once stretched all the way down to the Mohave Basin, south of modern-day Las Vegas, have been lost. The effects of a changing climate on rainfall, insect habitats and fungal diseases are not easy to predict, but could play a part in Methuselah's eventual demise – whether that is in a decade, a century or another millennium.

We humans have created for ourselves a climate conundrum. Ironically, it is only by being human, by doing what we do best – solving scientific problems and carrying out great feats of engineering – that we will find the solutions. We will not secure our children's place on this planet simply by filling old oil wells with carbon dioxide or hiding it all at the bottom of the sea. We must also recognize the value in the carbon stores that have built up naturally over thousands of years and fight hard to protect them. But all of this will come to nothing if we cannot find a way to power our future without fossil fuels. Our dependence on carbon-containing energy sources is what has led us into this mess and our legacy to our children should be the means to live without them.

← – –

Neptune grass (*Posidonia oceanic*) is a species of seagrass found only in the Mediterranean Sea, where it is declining due to pollution. Seagrasses globally are responsible for locking away around ten per cent of the organic carbon held by the oceans. Scientists estimate that their loss, at current rates, releases a million tonnes of carbon each year.

INSIDE EXPERT:
WHAT HAPPENS IF IT GETS TOO HOT?

"If we continue to rely mostly on fossil fuel, then it's going to be almost impossible to avoid two degrees [Celsius] of warming and there are a number of negative impacts that we expect already for a two degree temperature rise. As a climate scientist, I do not use words such as "dangerous" or "catastrophic". My work is to predict what would happen with climate for two or three degrees and my colleagues' task is to study the impacts of this climate change on different aspects of human life and natural ecosystems. After all, it is for people to decide whether these impacts will be catastrophic for them or not. So, for example, we are concerned about sea level rise because global warming could lead to complete melting of the Greenland ice sheet, which will cause the sea level to rise by 7 metres [23 feet]. For a two degree temperature rise, it will take a very long time, but the higher the temperature, the faster it will melt. There is also a danger of more extreme weather events and

we see this happening already. Quite a substantial number of ecosystems will be threatened because climate zones will shift and ecosystems will not have enough time to adjust. In terms of avoiding the worst impacts of climate change, I would say that I'm moderately optimistic, but in respect of avoiding two degrees of global warming, I'm rather pessimistic. Some technological revolution could happen tomorrow and we could forget about fossil fuels, but if that does not happen, I see no intentions of nations or politicians to do much about this. However, I think we can definitely stop somewhere between two and three degrees."

Andrey Ganopolski,
climate change expert,
Potsdam Institute for Climate Impact Research,
Potsdam, Germany

10

HOW DO WE GET MORE ENERGY FROM THE SUN?

I The grass in your garden does something incredible every single day: it soaks up the energy in sunlight and uses it to make food – a process known as photosynthesis. All green plants are capable of this and they do it not just to supply the occasional snack but to provide all of their sustenance. While we hunt and forage for food – albeit nowadays in the supermarket aisles – plants simply have to bask in the sunlight, using the power that sunlight provides to rearrange water and carbon dioxide into a chemical fuel, sugar, that they can use day to day. Of course, few of us would choose a drip-feed of sugar over the opportunity to select from a hundred different varieties of breakfast cereal, just to avoid moving from the sofa. But there is another reason we would like to copy what plants do: we need energy, and lots of it. What if we could power our kettles, cars and computers using the Sun's rays without clogging up the atmosphere with the dirty gases that come from burning oil and gas? Every hour of every day, the Sun is beaming more energy at the Earth than we get from fossil fuels in an entire year. If we humans could do what plants do and photosynthesize, converting even some of that energy into fuel, life would be far simpler. In 1912 the Italian chemist Giacomo Ciamician, one of the early pioneers of solar power, published his vision of a future where smokestacks were replaced by "forests of glass tubes" in which clean fuels were created using the same photosynthetic reactions as plants. The modern equivalent of Ciamician's glass forest is the artificial leaf, in which we will unlock the secrets of plants to turn sunlight into fuel.

To do what plants do is no mean feat. They may be brainless, but they have still had several hundred million years of evolution longer than us to refine the art of extracting energy from sunlight. The molecular manipulation that we've been trying to mimic takes place in reaction centres within the leaf where, in a chain of events that remains to this day somewhat mysterious, plant machinery uses packets of light energy called photons to split water molecules into their constituent components – hydrogen and oxygen – at an astonishing rate, carrying out up to 400 conversions every second (*see* BIG DISCOVERY: A CLOSER LOOK AT LEAF MACHINERY... IN BACTERIA). In doing so, it rips the charge-carrying electrons from hydrogen and sets up a chain of chemical reactions that eventually produce sugar. The part of this process that interests energy researchers is the water splitting bit, and what comes out of it: hydrogen. In a hydrogen fuel cell, energy is generated by reuniting hydrogen with oxygen to make pure, clean water as a waste product. Carbon is not involved in any part of this process. So hydrogen is just what we have been looking for – a clean fuel that has the potential to revolutionize twenty-first century living.

But why go to all the trouble of making an explosive gas – hydrogen – to use as a fuel when there is a more direct way to get power from the Sun? Staring at arrays of solar cells on rooftops, you might think that we have already worked out how to extract the energy from sunlight. In fact, some cheaper, lower-power solar panels use coloured dyes instead of silicon to absorb the Sun's rays, in the same way that a leaf uses the green pigment chlorophyll. These dye-based solar cells were invented in the 1980s by a Swiss scientist, Michael Grätzel, and are known as Grätzel cells. However, the big problem with all types of solar power is efficiency – or, rather, the lack of it. Unfortunately, even in the best solar panels currently available, most of the light hitting the panels is not converted into useful energy (usually in the form of electricity).

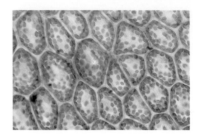

On the other hand, woeful inefficiency is by no means a reason to stop making solar panels, especially since making hydrogen with an artificial leaf is going to suffer from a similar problem, at least for the foreseeable future. Even a natural plant leaf can turn only about 1 per cent of the light energy that hits it into sugary fuel. Remember, though, to provide enough power for the entire planet we need only a fraction of the energy that is pouring through the clouds, so making the best solar panels possible (*see* INSIDE EXPERT: HOW DO WE MAKE BETTER SOLAR PANELS?), and making a lot of them, would go a long way toward solving our energy problems.

The photovoltaics that power the Mars exploration rovers are multi-layered and feed multiple electrical "junctions", allowing them to extract more energy from sunlight than a standard solar panel. At their peak, the wing-like panels supplied up to 900 watt-hours of energy per martian day.

Each leaf cell is packed with the green pigment chlorophyll, the substance that captures the energy in sunlight so that plants can make their sugary fuel. Small round packets called chloroplasts, which contain the chlorophyll, are visible under an ordinary light microscope.

← – –

There is another reason why solar power is not the whole solution: our addiction to fast, fuel-hungry cars. If the annual fuel consumption of the United States was represented by 10 barrels of oil, five of them would be used by cars and one by planes. You might be able to boil your kettle and run your computer on solar energy, but can you strap a solar panel to your car? And how long after driving off into the sunset will the battery wane, leaving you stranded till sunrise? As far as many scientists are concerned, what we need is a fuel; something that can be stored, moved around and used at any time of the day or night.

This is where hydrogen comes in. There is a certain attraction about the hypothetical hydrogen economy, which lies partly in the way that it seems to fit seamlessly into our oil-orientated vision of how fuel should be distributed. While charging your electric car (even one supplemented by the solar power at your eco-friendly house) has its advantages, it is perhaps easier to imagine visiting a hydrogen fuel station: a tanker delivers fuel to the filling station – just like with petrol; you visit the filling station to fill up your car – just like with petrol; and when fuel runs low, you stop at another filling station to pick up more – again, just like with petrol.

All we need to do is find a good source of hydrogen. However, that's where the problem lies: making pure hydrogen is more energy intensive and fossil-fuel hungry than is strictly permissible in our dream, zero-carbon scenario. Which brings us back to how plants do it and to the creation of the artificial leaf. If, like a humble leaf, we could take a water molecule and separate the two hydrogen atoms in it away from their oxygen using only sunlight, we would have a rich source of hydrogen. The tricky part does not lie in separating hydrogen from oxygen – we have been able to do that since the late eighteenth century, when a Dutch doctor by the name of Johan Rudolph Deiman and his friend Adriaan Paets van Troostwijk worked out how to split water by passing an electric current through it. No, the difficult part is doing it without using electricity (as plugging your clean fuel-generating device into an electric socket fed by fossil fuels rather defeats the point).

THE ARTIFICIAL LEAF

So how do we get the hydrogen out of water in the same way that plants do? In 1998, John Turner, a scientist working for the US government's National Renewable Energy Laboratory in Golden, Colorado, took a step in the right direction. He knew the key to solar water-splitting lay in bonding light-harvesting materials similar to those used in solar panels to other materials that would do the water-splitting. Turner made what is often referred to as the first artificial leaf, essentially a solar cell that split water when light fell on it. Its efficiency was decent, at 12 per cent – so much so that its unveiling caused quite a stir in

MIT professor, Daniel Nocera's "artificial leaf" is made of metal. But like a real leaf, it converts the energy in sunlight into a chemical fuel. It uses the energy to split water into oxygen and hydrogen, which can be used to feed a fuel cell in a hydrogen-powered car.

← - -

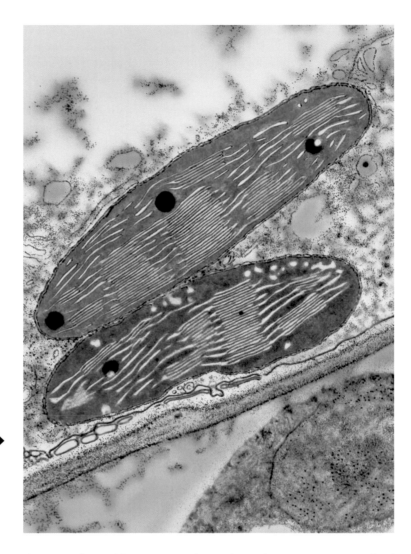

Close up on chloroplasts. Sliced lengthways, these chloroplasts from the leaf of the pea plant *Pisum sativum*. appear to contain stacks of flat, yellow pancakes. These are the grana, containing the light-trapping chlorophyll pigments.

the scientific world. But Turner knew this early attempt at an artificial leaf was nowhere near good enough in practical terms. For a start, the materials that drove the water-splitting reaction – the catalysts – were rare metals that would be far too expensive to use as the basis for any commercially viable technology. Worse still, Turner's water splitter started to disintegrate after just a day. The problem was one that others encountered too; the better a material was at splitting water, the faster it seemed to fall apart in sunlight.

Over a decade later, a team of scientists led by Daniel Nocera at the Massachusetts Institute of Technology found an ingenious way to circumvent the problem when they presented the first practical design for an artificial leaf. They used a self-repairing catalyst that put itself back together as fast as it broke down. Inspired by Ciamician's forest of glass tubes and seeking an inexpensive fuel source for the poorest parts of Africa, Nocera compromised on efficiency to focus on splitting water using cheap, widely available metals as catalysts. In 2011, he demonstrated his prototype device. At about the size and shape of a credit card, it looked completely unremarkable, but when dropped into a small container of water sitting in a shaft of light, bubbles started to stream from each side – hydrogen and oxygen. It was no coincidence that the fine structure of the catalyst Nocera used resembled its equivalent molecules in plants. But while Nocera's device does the job cheaply, he will have to wait to see how it fares after a few months, or years. A solar panel can continue to produce energy for decades, but can an artificial leaf? There is also the question of efficiency – at under 3 per cent efficient, Nocera's device still has plenty of room for improvement.

With artificial photosynthesis finally coming of age, the idea of filling up your car at a hydrogen station fuelled by water and sunlight alone

Large solar panel arrays have been set up to help power research facilities, and army and naval bases, but many also supply electricity to national grids. Power generation depends on insolation – how much sunlight reaches a given area over a given time. Insolation values are high in Australia, for example, but low in Norway.

Shown under construction, the Joint European Torus (JET) in Culham, UK, is the world's largest and most powerful tokamak – a nuclear fusion machine. While today's nuclear power stations generate energy by splitting atoms (fission), a tokamak generates energy by squeezing atoms together (fusion) and without producing any highly radioactive nuclear waste.

INSIDE EXPERT: HOW DO WE MAKE BETTER SOLAR PANELS?

"Today, most solar panels are made from silicon – a semiconducting material that absorbs solar radiation and converts it into electrons. To get a current, you need something called an electrical junction, which passes the electrons into a circuit. Now with a single junction you can only absorb the light from one part of the spectrum, so even the best laboratory solar cells made this way are only about 25 per cent efficient. But there are some very high efficiency cells that get over 40 per cent by using multiple electrical junctions, fed by multiple layers of semiconducting materials capturing light from different parts of the spectrum. These are complex structures built for powering satellites and they're expensive. But there's interest in using those in solar concentrators – you focus high concentrations of solar radiation on a small area and you cut your costs because you're only using a small amount of material. In the future, I think we'll see a diversification of solar technologies so that we'll be getting energy not just from solar panels, but from these solar concentrators and from fuels generated directly by the Sun. The other exciting thing about solar cells is you can put them into buildings. We've all seen the panels bolted on to houses, but you can imagine buildings of the future where the building fabric itself is actually generating electricity. In my research I work on thin-film materials, which can be made into panels in the same way as silicon, but they could be laminated on to roof tiles or made into windows. Think of factory roofs, retail parks – if we just used our buildings, we could generate a really significant proportion of our electricity from the Sun. I think that's a very exciting vision for the future. Then if you look to the end of the century, where will the technology be? Seventy-five per cent of our electricity generated from the Sun? It's not unrealistic."

Professor Stuart Irvine, Director of Centre for Solar Energy Research, Glyndŵr University, UK

BIG DISCOVERY:
A CLOSER LOOK AT LEAF MACHINERY...
IN BACTERIA

Understanding how natural photosynthesis works is a real challenge, not least because some of the key components in the leaf's light-harvesting machinery are stuck in membranes inside cells, making them almost impossible to get out in one piece. But in the 1980s, three German scientists uncovered the hub of the whole process, the photosynthetic reaction centre. They did it by turning their attention from plants to bacteria, probing the simpler structures in a purple bacterium called *Rhodopseudomonas viridis*, which photosynthesizes in a similar way to plants.

"For a long time it has been impossible to prepare membrane-bound proteins in a form allowing the determination of the detailed structure in three dimensions. Before 1984, there were only rather fuzzy structural pictures available for a few membrane proteins... But the situation had actually drastically changed in 1982, when Hartmut Michel...succeeded in preparing highly ordered crystals of a photosynthetic reaction centre from a bacterium. With these crystals he could in the period 1982–1985, in collaboration with Johann Deisenhofer and Robert Huber, determine the structure of the reaction centre in atomic detail...[leading] to a giant leap in our understanding of fundamental reactions in photosynthesis, the most important chemical reaction in the biosphere of our Earth."

From the presentation speech for the 1988 Nobel Prize in Chemistry awarded to Johann Deisenhofer, Robert Huber and Hartmut Michel "for the determination of the three-dimensional structure of a photosynthetic reaction centre".

no longer seems quite so farfetched. But it is not the end of the story for hydrogen. Another promising alternative energy source is fusion – forcing hydrogen atoms together to make helium and knocking out tiny particles called neutrons in a reaction that releases immense amounts of energy. Fusion does not use sunlight, but instead is based on the processes that generate energy in the centre of the Sun itself. Some scientists are convinced that fusion can provide the energy we need to power our future, but there are huge technical problems to be overcome first. To make fusion happen on Earth, you have to create conditions similar to those that drive fusion in stars. In other words, you have to make an environment that is as hot as the centre of the Sun; or, to be completely accurate, *hotter* than the centre of the Sun, in order to compensate for the relatively low pressures that can be achieved on Earth. The Joint European Torus (JET) fusion experiment in Oxfordshire, UK, operates at a staggering 150 million degrees – about ten times hotter than the centre of the Sun – and is powered by multiple heating systems using millions of watts (megawatts) of electricity. To be worthwhile, any viable fusion technology would have to generate a healthy energy profit to cover its colossal energy cost. With this in mind, the production of 16 megawatts of fusion power by the JET machine in 1997 may sound less than impressive, but it is a feat that remains unmatched. After some downtime from 2009 to 2011 for equipment upgrades, JET is due to resume fusion testing and its eventual aim is to produce ten times the energy it consumes.

Our future on this planet depends on finding clean sources of energy – and undoubtedly we will need more than one – to take the place of fossil fuels. In doing so, we could do worse than look to nature for inspiration. Whether we are mimicking the processes that occur in leaves or reproducing the reactions inside our closest star, we are doing only what nature in all its wisdom has already deemed possible.

Plants get their characteristic colour from the leaf pigment chlorophyll, which absorbs the red and blue-violet parts of sunlight but reflects the green.

← - -

The core of the JET nuclear fusion experiment is built to hold extremely hot plasma. It is maintained and repaired by a robot – the Mascot "Slave" unit – with an artificial hand (left of this image) that can be controlled remotely from outside the machine.   - - →

11

WHAT'S SO WEIRD ABOUT PRIME NUMBERS?

I took mathematician Leonhard Euler seven weeks to make the journey from Basel in northern Switzerland to the historic city of St Petersburg. The son of a clergyman, the 20-year-old was heading for the Russian Academy of Sciences, which had been established two years earlier in 1724 by Tsar Peter the Great. In travelling east, Euler was following in the footsteps of several of his fellow countrymen who had flocked to join the newly formed institution. Surrounded by like-minded contemporaries and a vast library, partly drawn from the Tsar's own private book collection, Euler began to tackle some of the great mathematical puzzles of the day, but his passion remained with the area of mathematics that had entranced him since childhood: prime numbers.

As many schoolchildren can recite, a prime number is any number that can only be divided by itself and one: 2, 3, 5, 7, 11, 13, 17, etc. (1 is not a prime number because it can be divided by only one number – itself.) These numbers are often referred to as the building blocks of mathematics because, around 2,000 years before Euler's trek to St Petersburg, the ancient Greek mathematician Euclid had shown that any non-prime number can be made by multiplying prime numbers together. Today, prime numbers are even more important – much of modern mathematics is based on them and, on a practical, everyday level, they form the basis of the system keeping your credit card details out of the hands of unscrupulous internet thieves.

EUCLID TO EULER

Since the time of Euclid, mathematicians have searched for a pattern in primes, for some order to the basis of their craft. They hoped to find a formula that could produce any number in the prime number sequence. For some number sequences, such a formula is relatively straightforward to find. Take the following sequence: 1, 4, 9, 16, 25... These are the "square" numbers – each number is obtained by multiplying its position in the list by itself ("squaring it"). So, 1 x 1 = 1; 2 x 2 = 4; 3 x 3 = 9 and so on. If you want to work out the sixth number in the list, just multiply 6 by itself (6^2 = 6 x 6 = 36). Working out the 108th number is just as easy – it's 11,664 (108 x 108). The formula for finding any square number is fairly simple. Could the same be done with the primes?

This question brings us back to Euler. By the time of his second stint in St Petersburg (he'd fled to Berlin for 25 years to escape the political turmoil in Russia), Euler had hit upon a way to generate prime numbers. His formula was $x^2 + x + 41$. Start putting numbers in place of x and you'll get a prime: with x = 1 you get (1 x 1) + 1 + 41 = 43, which is a prime number; with x = 2 you get (2 x 2) + 2 + 41 = 47, also a prime number. Euler's formula continues to generate prime numbers if it is fed with every number up to, but not including, 40. But while his formula hints at a possible underlying pattern within the primes, it can generate only a small portion of the list. Euler's fascination with prime numbers continued throughout his life, but he died of a brain haemorrhage in St Petersburg in 1783 without finding a universal formula for all prime numbers.

German mathematician Georg Friedrich Bernhard Riemann (1826–1866). He noticed a relationship between the zeta function and the distribution of the prime numbers. Proving a related hypothesis – the Riemann hypothesis – remains one of the biggest outstanding problems in modern mathematics.

- - →

THE RIEMANN HYPOTHESIS

By 1859, the German mathematician Bernhard Riemann provided a possible way in. Working at the University of Göttingen, Riemann was studying an area of mathematics known as the zeta function, first studied by Euler himself almost a century earlier. A function is

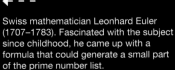

← - -

Swiss mathematician Leonhard Euler (1707–1783). Fascinated with the subject since childhood, he came up with a formula that could generate a small part of the prime number list.

just a way of doing something to a number – input a number into a function and you get another number as an output. Euler's $x^2 + x + 41$ is just such a function (although not a zeta function). Riemann inputted some numbers into the zeta function that Euler hadn't tried and became particularly interested in those that gave an output of zero – they seemed to form a pattern. When he plotted them on a graph, the ones he'd found all sat on a straight line. What's more, Riemann could also see a relationship between the inputs that made the zeta function output a zero and the prime numbers. If there was a hidden order in the zeros, perhaps there was an order in the apparent chaos of the primes?

Riemann guessed that, with a few exceptions he already knew about, every other possible input that gives a zero (including all the ones he hadn't found) also lies on that same straight line – the pattern in the zeros never deviates. This is called the Riemann hypothesis. If this hunch is true, it could unveil the mystery surrounding a possible pattern

in the prime numbers. Unfortunately, we don't know if it is. It is possible that Riemann himself had proof but, keen not to submit any work he deemed incomplete, he never published on prime numbers again. As the Austro-Prussian War reached Göttingen in 1866, Riemann fled to Italy, leaving many of his papers behind. His housekeeper destroyed much of his unpublished work. Later that year, Riemann died of tuberculosis, aged just 39.

By the end of the nineteenth century, proof of the Riemann hypothesis remained one of the most sought-after results in all of mathematics. Mathematicians longed to find order in the seemingly chaotic way the primes – the bricks from which they'd built their mathematical house – were distributed. As the nineteenth century passed into the twentieth, another German mathematician, David Hilbert, gave a lecture casting his eye over the future mathematical landscape. In his address to a Parisian audience in the summer of 1900 he presented what he considered to be 10 of the most pressing problems facing mathematicians in the new century. He later published the list, adding 13 additional problems. The full collection, known as Hilbert's problems, had proving the Riemann hypothesis at number eight. In the decades that followed, many of Hilbert's problems succumbed to mathematical advances. Yet despite efforts by some of the twentieth century's best mathematicians – including the Nazi-code-breaking Alan Turing – proof of the Riemann hypothesis remained an outstanding problem in mathematics by the dawn of the present century.

In the meantime, many other advances have been made in mathematics based on the assumption that the Riemann hypothesis would one day be proved true. This is not as crazy as it may sound – modern computers have shown that billions of inputs that make the Riemann zeta function equal to zero do indeed all lie on the straight line. However, finding just one that doesn't would break the very foundations that a lot of modern mathematics is built on. Such is the importance of a proof of the Riemann hypothesis, the Clay Mathematics Institute based in Cambridge, Massachusetts, has offered a prize of $1 million to anyone who can provide it (*see* BIG DISCOVERY: PROVING THE RIEMANN HYPOTHESIS). The money forms part of their seven Millennium Prize Problems – analogous to Hilbert's problems, they were announced in 2000 as the calendar rolled into another new century. One has already been solved (the Poincaré conjecture, which concerns three-dimensional spheres and was solved in 2002–2003), but a solution to any of the remaining six would make you an instant millionaire. So a pattern in the primes clearly matters to mathematicians but it has also become increasingly important to governments, businesses and consumers thanks to a new technology that sprang up towards the end of the millennium: the internet.

German mathematician David Hilbert (1862–1943). At the turn of the twentieth century he published a list of 23 of the biggest unanswered questions in mathematics. Number 8 – proof of the Riemann hypothesis – still alludes mathematicians to this day.

BIG DISCOVERY:
PROVING THE RIEMANN HYPOTHESIS

A "bucket list" – a set of goals to accomplish before you die – has become a popular phrase in recent years. There was even a 2007 film of the same name starring Morgan Freeman and Jack Nicholson. And, for many mathematicians, proving the Riemann hypothesis – the discovery that there might be a hidden order within prime numbers – sits right at the top of theirs. Long before the Clay Mathematics Institute offered a $1 million prize for a proof in 2000, early twentieth century British mathematician G.H. Hardy wrote the following New Year's resolutions on the back of a postcard to a friend:

1. To prove the Riemann hypothesis.
2. To make a brilliant play in a crucial cricket match.
3. To prove the non-existence of God.
4. To be the first man atop Mount Everest.
5. To be proclaimed the first president of the U.S.S.R., Great Britain, and Germany, and
6. To murder Mussolini.

Hardy worked intensely on the Riemann hypothesis, but never came up with a proof. Such a proof would have solved one of the 23 problems German mathematician David Hilbert tasked mathematicians with resolving in 1900. Hilbert is even quoted as saying, "If I were to awaken after having slept for a thousand years, my first question would be: has the Riemann hypothesis been proven?" He died in 1943 – if he were to come back to life today, he would surely be disappointed with the answer he'd be given. In fact, proof of the Riemann hypothesis, first put forward in 1859, remains perhaps the most elusive entity in all of mathematics. Modern Italian mathematician Enrico Bombieri has called the issue the "central problem of pure mathematics". Of course, proving that the Riemann hypothesis is false – that is, that there is no hidden order in the prime numbers – would be equally significant, not least for the implications it might have for the security of online and other coded transactions that currently rely on primes.

CODE-MAKERS

The internet has revolutionized many aspects of modern life, not least the fact that more money is now spent online than in physical shops. With all these credit card details flying around cyberspace, there needs to be a way of keeping banking details secure, a way of encrypting the data so it can't be intercepted by would-be fraudsters. This art of cryptography is nothing new – one of the earliest and most basic ways to encrypt a message is to use a Caesar cipher, named after Julius Caesar, who used the method in his correspondence. In this method, you simply change each letter by a set amount; for example, changing "please don't read this" by two letters gives "rngcug fqp'v tgcf vjku". Someone at the other end, knowing what you've done, can turn the message back into its original form. The same can be done with the string of numbers on a credit card. But this form of encryption is far too simple to break. There is also another problem: at some point you have to send the key – the secret of how to decipher the message. Imagine that, rather than encoding the message, you put it in a locked box. At some point you need to send the recipient the key to that box. If the key is intercepted, then the interceptor can read your messages and all future messages sent the same way. It is much safer to have one key that locks the box and a different key to unlock it, and prime numbers offer a way to do just that.

The main way information is currently encrypted online is called public-key (PK) encryption, which uses a pair of mathematically related keys, one that is made public and another that is kept a secret. If I want you to send me a message, then I give you my public key and ask you to use it to encode a message. The only way the message can be decoded, however, is with my private key. So anyone can encode a message using my public key, but only I can read it once it has been encrypted. Your web browser scrambles your credit card details using a website's public key, but only the website can unscramble it because their private key is known only to them. The keys are generated by multiplying two very large prime numbers together to get a bigger number. The big number forms the basis of the public key; the two original prime numbers help form the private key. This works because, even though the big number is widely available, it is very hard to work out what the two prime numbers were. The method for turning the prime numbers into the two keys is known as the RSA algorithm, named after Ron Rivest, Adi Shamir and Leonard Adleman, who came up with the technique at the Massachusetts Institute of Technology in 1977. In 1997, however, the UK's Government Communications Headquarters (GCHQ) announced that the same technique had been devised by British mathematician Clifford Cocks and his colleagues in 1973 but had remained classified. This just goes to show how important

The internet has revolutionized modern life. More money is now spent online than in physical shops and prime numbers currently play a crucial role in encrypting sensitive banking information online.

prime numbers are to security – for both the military and the internet.

It is not that RSA encryption cannot be broken, it's just that to crack the code you need to find the original prime numbers and that currently takes an impractically long time. One of the reasons for this is the seemingly random nature of the primes, the very thing that Riemann looked to tame. To try and demonstrate the security of their technique, Rivest, Shamir and Adleman published a 129 digit number (now known as RSA-129) in the pages of Scientific American magazine in August 1977. They offered a $100 prize for anyone who could tell them the original two prime numbers they'd multiplied together. The answer did arrive eventually, but not until 1994. The team who eventually came up with the answer used the internet to harness the spare computing power of around 600 volunteers worldwide to crack the code in eight months. The current record for the highest RSA number split into its two primes is RSA-768, which was cracked in 2009. The prime numbers most widely used in current PK encryption are each at least 1,024 digits long.

Credit cards details are protected online using a technique called public-key (PK) encryption. The RSA algorithm converts two very large prime numbers into two keys that can encode and decode your payment information without it falling into the hands of fraudsters.

But as computers continue to get faster (*see* QUESTION 12: CAN COMPUTERS KEEP GETTING FASTER?), so does their ability to crack the RSA codes. This means that as more and more RSA numbers fall by the wayside, higher and higher prime numbers are needed to keep the internet secure. But without a sure knowledge of any pattern in the prime numbers, how can you guarantee a continued supply of bigger and bigger primes? How do you know the next prime in the sequence? After all, you have to be sure that each of these enormous numbers cannot be divided by any other number than itself and one. Luckily, mathematicians have developed tests that can quickly prove whether a number is a prime or not. Crucially, however, many of these tests rely on the Riemann hypothesis being true.

So assuming the Riemann hypothesis is true helps to engineer the prime numbers that keep internet transactions secure. But these RSA numbers also take so long to crack because of the seemingly random way the primes are spread out. As the Riemann hypothesis seeks to lift the veil on their chaotic nature, some mathematicians have suggested that a successful proof may also provide insights into how to break the RSA codes much faster. If they're right, the whole of the internet's current security infrastructure could be threatened. It is no wonder, then, that considerable attention is paid to the seemingly abstract work going on in mathematics departments across the globe.

Some of these mathematicians are working on alternative encryption methods that don't rely on primes. But, for now, it is that boyhood obsession of Euler that is keeping your credit card details out of the hands of criminals. Solving the Riemann hypothesis – finding a true pattern in the apparent weirdness of the primes – might change all that.

CAN COMPUTERS KEEP GETTING FASTER?

he morning in April 2005, British engineer David Clark found himself pulling up floorboards. This was no home improvement project. Clark, a hoarder, was trying to find one particular piece of memorabilia that would make his fortune – or at least fund his daughter's wedding. The item in question was an old copy of the trade magazine *Electronics*. Prompted by a news alert from one of his favourite technology websites, Clark had learned that morning that a volume 38 number 8 issue of the magazine from 1965 could fetch $10,000. Consumed by the idea that a copy might be in his possession, he took the morning off work to hunt through an unofficial archive hidden under the floorboards of his home. The first pile he pulled up brought no joy, but when he unwrapped the second pile the magazine was sitting right on the top, in mint condition. A week later, he passed it over to computer giant Intel in return for the promised $10,000 prize. Meanwhile, another volume 38 number 8 issue of the magazine had mysteriously disappeared from the engineering library at the University of Illinois. So what was it about the 40-year-old journal that made it so highly sought after?

The object of Intel's interest was an article by the company's founder, engineer Gordon Moore, billed in the piece as "one of the new breed of electronic engineers, schooled in the physical sciences rather than in electronics." In the article, Moore casually made a prediction that would become one of the guiding principles of computing. Moore's law, as it is now called, states that the number of components on a computer chip doubles every two years. What Moore actually wrote in that prized edition of *Electronics* was that the number of components would double every year for at least 10 years. A decade on, speaking about the progress of chip complexity at a meeting of the Institute of Electrical and Electronic Engineers, he admitted it would not be possible to sustain this rate and revised his "law" to account for biennial doubling. At that time, he expected the numbers of components on a chip – governing computing speed and memory – to have started levelling off by the mid-1980s, as "circuit and device cleverness" approached an upper limit. On the contrary, chip complexity kept increasing; engineers went on making the electronic components smaller and packing them tighter. Since the inception of Moore's law, the number they could cram on to a computer chip soared, from fewer than 100 in 1965 to a billion by 2005. It seemed Moore's law knew no bounds.

In the space of half a century, computers have shrunk from machines the size of entire rooms to chips small enough to fit inside the phones in the palms of our hands. In retrospect, it seems unbelievable that the guidance system that carried Neil Armstrong and Buzz Aldrin to the Moon in 1969 had less computing power than a modern electronic toaster. Which leads us to ponder: if we can fly to the Moon with the computing equivalent of a toaster, why do we need more powerful computers? Every mobile phone, tablet, digital watch, MP3 player, washing machine and car already has a miniature built-in computer. Perhaps it is for our pursuit of art and science that we demand this greater processing power, for everything from sophisticated special effects in the movies, to satellite mapping and drug design. Or perhaps it is as simple as saving time and money.

Whatever the reason, it has become imperative that Moore's law holds true. In April 2012, Intel launched its "Ivy Bridge" chip, manufactured with components that are a minuscule 22 nanometres (22 millionths of a millimetre, or about 8.7 ten-millionths of an inch) in size and operate at a scale smaller than dust mites, bacteria and even our own chromosomes. The company boasted of being able to fit more than six million transistors – the miniature electrical devices that switch and amplify electrical signals in an integrated circuit or silicon chip – into a space the size of a full stop. Circuit and device cleverness have exceeded even Moore's expectations. There is, however, a limit to the number of transistors that engineers can physically fit on to a chip using traditional silicon technology without making the chip itself dramatically bigger. That limit exists at the level of the fundamental building block of matter: the atom. If we follow Moore's law to the letter, by the year 2020, the fastest computers will be running on chips using millions of billions of atom-sized transistors. And if an electronic component fashioned from an individual atom sounds like science fantasy, it is far from it. In fact, it already exists. In 2012, after 10 years of experimentation, a team of US and Australian scientists unveiled their single-atom transistor – one phosphorus atom sitting on top of a slice of silicon.

Gordon Moore co-founded the Intel computer corporation in 1968. Previously he was known as one of the "traitorous eight" who left the Shockley Semiconductor Laboratory in 1957 to set up Fairchild Semiconductor, where the first silicon computer chip was created.

TINY TRANSISTORS

In terms of Moore's law, these atomic-scale transistors put us a few years ahead of schedule. There is a problem, though. The smaller chip engineers make their transistors, and the tighter they pack them, the more difficult it is to get them all working together. At any one time, a chip might be operating with the power supply to millions of its components shut off in order to avoid overheating and malfunctioning. According to computer scientists, who refer to the problem as "dark silicon", around a fifth of the transistors on a chip would need to be turned off if they were the size of those on Intel's Ivy Bridge and around half if they measured 8 nanometres (8 millionths of a millimetre, or about 3 ten-millionths of an inch). Writing in 2011, in a paper devoted to the issue of dark silicon, researchers from Microsoft and the Universities of Texas at Austin, Washington and Wisconsin-Madison predicted the industry would hit a wall representing the end of Moore's law within just five years. The onus, they wrote, would be on computer architects to deliver performance and efficiency gains from existing chip technology.

It looks like we could be reaching the end of an era. But atomic-level computer engineering does not have to use conventional methods; there

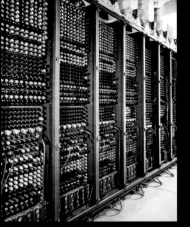

Constructed at the University of Pennsylvania in the 1940s, the Electronic Numerical Integrator and Computer (ENIAC) filled a room 9 by 15 metres (30 by 50 feet). It was a giant calculator used to work out the trajectories for firing weapons, meaning the time taken to perform a calculation that would have taken 12 hours by hand was reduced to under a minute.

are other ways to make a computer. The atomic scale – the quantum scale – might be the beginning of a new and more exciting age of computing. It is at the quantum scale that things start to get really interesting, because not everything works in the same way as it does at larger scales. Weird things happen, like particles being in two places at once or behaving in different ways at the same time. A quantum computer, based on quantum bits of information, would be very different indeed to a regular desktop computer. To understand what's weird about this potential new form of computing, you have to understand how information has been stored in chips since the middle of the last century. In a standard silicon chip, each "bit" of data or information can have one of two values: 0 or 1. This is why computers are said to work in binary, meaning two parts. Which value a bit of data actually holds is determined by the configuration of a tiny switch, a transistor. But in a quantum system, particles can be in two states at the same time, so a quantum bit (called a qubit) can have three possible values: 0, 1 and 01. Yes, 01, where a qubit holds both 0 and 1 values at the same time. This may not seem like such an advantage until we realize what happens when there are more bits. For example, while a normal data system can store only three binary digits in one configuration (say, 110), a quantum system could store all eight possible variations of 0s and 1s in a particular three digit combination. In theory, not only could a quantum computer store all this extra information, it could also do simultaneous calculations with it, making such a computer

The first Apple computer – Apple I – was a basic circuit board devised by Steve Wozniak, Steve Jobs and Ron Wayne in 1976. Users added their own display units and keyboards, giving each computer a truly personalized look.

– – →

ultra ultra-fast. As if all that is not weird enough, one interpretation of the theory says that quantum computers are so fast because they perform calculations in multiple universes (*see* QUESTION 8: ARE THERE OTHER UNIVERSES?) at the same time.

Quantum computers, then, are a strange and enticing prospect, but many have doubts about their feasibility. While experiments have demonstrated that a quantum computer could work in principle, scientists have not yet managed to build one that handles more than a few qubits of data. As the key to the quantum computer's power is running multiple computations in parallel, the qubits – made in many cases from individual particles – have to be linked, or as a quantum physicist would say, "entangled". Due to another aspect of quantum bizarreness, it is possible for quantum objects to be entangled such that the behaviour of one affects the behaviour of another, without them even touching each other – what Albert Einstein referred to as "spooky action at a distance". Unfortunately, however, entangling lots of qubits turns out to be much more complicated than anyone first imagined. It is hard enough building a system that can handle three or four qubits let alone one containing hundreds, thousands or millions. One issue is that qubits are slippery. It takes only a fraction of a second for them to fall out of the weird superposition of states that allows them to hold two values at the same time, so the lifetime of a quantum bit may be only a matter of microseconds.

The 1s and 0s of binary represent an on-off state or yes-no choice. Although the first electronic computers were not invented until the twentieth century, binary code has been around since the seventeenth century, when English mathematician Eugene Paul Curtis adapted binary principles from an ancient Chinese text.

COMPUTING COMES TO LIFE

With so many challenges ahead for physics, quantum computing may not immediately be able to fill the void left by silicon when Moore's law finally fails. In the meantime, however, chemistry and biology might hold the key to greater computing power. One intriguing way of creating very small computer components is to build them out of the stuff of life: DNA. Like other molecules, strands of DNA can be used as the basis for bits of data. The molecule has its own code – in life's blueprint this relates to the sequence of linked parts in the DNA chain, but in a DNA computer this code could perform a similar function to binary 1s and 0s. As early as 1994, Californian computer scientist Leonard Adleman showed that it was possible to apply biological molecules to solve difficult problems. He used a test tube full of DNA to work out the answer to a version of the travelling salesman problem – a classic problem in computing that requires working out an optimal route to a final destination visiting certain cities along the way (*see* BIG DISCOVERY: BIOLOGICAL COMPUTERS SOLVE HARD PROBLEMS).

So DNA computers are good at getting the right answers, but will they ever be quick enough to challenge silicon? In 2011, another

BIG DISCOVERY:
BIOLOGICAL COMPUTERS SOLVE HARD PROBLEMS

"…the potential of molecular computation is impressive. What is not clear is whether such massive numbers of inexpensive operations can be productively used to solve real computational problems…Nonetheless, for certain intrinsically complex problems…where existing electronic computers are very inefficient…it is conceivable that molecular computation might compete with electronic computation in the near term."

From "Molecular computation of solutions to combinatorial problems", Leonard Adleman, published in the journal *Science*, 11 November 1994.

US computer scientist Leonard Adleman showed in 1994 that it was possible to set up a biochemical experiment to search for a solution to a problem in the same way as a computer algorithm. He used DNA to solve the travelling salesman problem – finding a route between a number of imaginary cities without ever visiting the same city twice – for seven cities. Each city was represented by a short piece of DNA with a particular code. Each road between two cities was represented by another piece of DNA containing code to match the departure city code at one end and the arrival city code at the other. Pieces of DNA stick naturally to other pieces with matching codes, so when all the pieces were mixed together roads and cities immediately started linking together to form "journeys". Adleman then used molecular techniques to filter out the linked pieces of DNA that did not contain the codes for all seven cities – the molecular equivalent of sacking any salesman who did not visit each city as instructed. The route could then be read from the order of the linked DNA pieces. Although Adleman's seven-city problem is relatively easy to solve, the complexity of the travelling salesman problem quickly multiplies with additional cities, eventually requiring a supercomputer to calculate the answer.

Californian scientist, Erik Winfree, and his team built a computing device known as a neural network, so called because it contains processing elements that mimic connections in the human brain. They used DNA to form the basis of the network and engineered it to identify patterns in new strands of DNA that they added to the mix. The newly added strands stuck to DNA within the network to generate what the scientists called input signals and triggered a cascade of biochemical reactions that would eventually lead to a visible output signal in the form of a colour change. Each new strand they added contained a code representing a response to a simple question answered by a scientist in the laboratory, so that by recognizing a certain combination of answers – a pattern in the strands – the network was able to "guess" at the scientist who gave them. By adding the extra strands, Winfree was essentially asking his biological computer: "Which scientist gave the following responses?" The computer recognized the patterns – it identified the right scientist – every time, but took eight hours to do so.

Because of its slow speed, DNA computing is more likely to find applications in the field of medicine, where biocompatibility is paramount. For example, DNA devices could be programmed to seek and destroy cancer cells. But biological computing also tests our preconceptions of what a computer should look like. What about trying to make a computer in a more conventional way but with unconventional materials? Silicon is the standard material because it is widely available, cheap and conducts a current at just the right level for electronics. Its conductivity can also be tweaked by adding small amounts of other elements. There are other materials that could do the same job, but they tend to be expensive. For years now, graphene

– a one-atom-thick sheet of carbon that can be peeled from pencil lead – has been high on the list of candidates for replacing silicon. It is a strong, flexible, mesh-structured material and a super-speedy electrical conductor. But there is such a thing as too conductive. While with silicon the current can be switched on and off, the same is not true with graphene – its wafer-thin sheets of carbon allow electrons to move across its surface unhindered. So scientists are searching for ways to tailor the structure and conductivity of graphene to make it semiconducting and suitable for use in transistors.

In 2015, Moore's law will be half a century old. Less a law than a self-fulfilling prophecy, it has driven computer engineers to make smaller and smaller components and cram more and more of them on to their chips, through sheer weight of expectation. Moore himself has admitted his "law" was nothing but a lucky guess – he saw a trend and extrapolated. It was his friend Carver Mead, an engineering professor, who dubbed this guess "Moore's law". Whatever it is – a guess, a law, a prophecy – it did not just predict the future, it shaped it. The computer industry has set itself the challenge of keeping up. Perhaps a new prophecy is precisely what is needed to drive forward a new era in computing, when we have decided what that might look like.

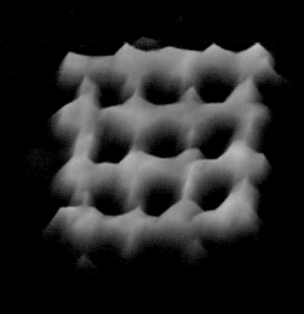

- - →

This waffle-like structure is a logic circuit made from DNA that uses light instead of electricity to transmit a signal. One of the advantages of DNA computing devices is that they can self-assemble, and in large

WHEN CAN I HAVE A ROBOT BUTLER?

…if every instrument could accomplish its own work, obeying or anticipating the will of others… chief workmen would not want servants, nor masters slaves." So said Aristotle, 2,500 years ago in his famous work *Politics.* Considering the tripods of Hephaestus – walking trays serving the blacksmith of the Greek gods and fictionalized by the poet Homer in the *Iliad* – Aristotle speculated that if such intelligent tools existed, we might achieve human equality by abolishing servants and slavery.

Human equality is a tall order, but the idea of an automaton that frees us from mundane work is core to any "robot" – when the writer Karel Čapek coined the term in his 1920 play *R.U.R. (Rossum's Universal Robots),* he derived it from the Czech word for servitude. Leonardo da Vinci is among the famous historical figures known to have dabbled with automated mechanical devices, but it wasn't until the 1940s and the advent of computing that things really took off. By 1961, US company General Motors had the first ever industrial robot, Unimate, assembling cars in one of its factories. Manufacturing hasn't looked back since.

Yet 60 years later, there's still something sci-fi about a domestic robot. Part of the reason is that we don't notice the many robots already all around us. Factory robots are out of sight for most people, as are the agricultural robots that milk cows. We don't think much about the driverless trains at airports and we hardly notice the virtual assistants that help sort our email. A robot butler is not out of the question, although it depends on what exactly you're asking for.

READY TO SERVE

What would you want a robot butler to do? And what would it need to do those things? If it's simply fetching and delivery, there are already plenty of robots that do that. If you've ever ordered from an online retailer, the chances are that your goods were brought to you via a Kiva. These squads of small, bright orange robots ferry items (or rather, the whole shelf unit containing the items) directly to the human packer. They also help to keep a massive warehouse in order – since it's all coordinated by a central computer, the system moves the most popular items closer to the packing station. For this to work, the robots need the right infrastructure in terms of compatible doors, elevators and furniture. It's not outlandish to imagine that a house of the future might be designed with such robots in mind. Building a robot that can cope with our current houses is another matter, although we do have Honda's ASIMO as proof-of-concept. The Japanese company revealed this, the world's first walking humanoid robot, back in 2000. At 130 centimetres (4.3 feet) tall and weighing 48 kilograms (106 pounds), the 2012 ASIMO can move at up to at 9 kilometres (5.6 miles) per hour, jog in circles, walk on uneven surfaces, hop and even climb stairs.

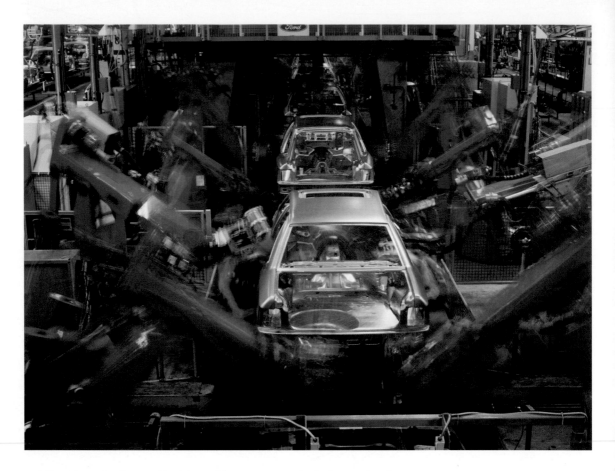

ASIMO's multi-fingered hands enable it to pour a drink from a bottle – it is even able to twist off the cap and handle a paper cup without squashing it. These dextrous hands benefit from touch and force sensors in the palms and fingers, which give feedback to the robot's processor and enable it to control each finger independently. Combined with object recognition technology based on its visual and tactile senses, this allows ASIMO to carry out its tasks and even to use sign language. It's a far cry from the stiff-jointed arms and claws of industrial robots, although it remains to be seen whether robot grippers can handle objects that change in size and shape, such as sheets, cushions and wires. Such challenges might be met by some out-of-the-box thinking. Researchers at Cornell University and the University of Chicago are working on a simple, passive universal gripper that's like a balloon filled with sand. The balloon is placed over an object and a vacuum is turned on, pumping air out of the balloon and causing it to conform to the shape of the object and grip it rigidly. To release the object, air is let into the balloon or, alternatively, a compressor can rapidly pump air into it to launch the object at speed.

The vast majority of cars and trucks in the developed world are produced by robots. Robots can be programmed to perform repeated operations with great precision, hence their widespread use in manufacturing. Their rates can be set and integrated to produce a continuous production line flow.

This has proved remarkably dextrous in trials, allowing a robot to sort nuts, bolts, balls and darts accurately into boxes – and to shoot basketball hoops.

SENSE AND SENSOR-BILITY

We've developed the technology for sensing almost anything in the environment: light, sound, pressure, temperature, movement, and even smell. Some of these sensors would be useful in helping a robot butler know where it is. In fact, some sophisticated technology is already in our living rooms: Microsoft's Kinect for the Xbox games console uses a combination of cameras and software to scan the whole room and build a 3D image, distinguishing what is there and what is a moving human. The device has been a boon to robotics researchers and enthusiasts, offering a sophisticated movement sensor off the shelf. Companies like the US-based iRobot have built the technology into robots such as Ava, a remote-controlled teleconferencing robot. Ava is interesting for several reasons, one of which is iRobot's partnership with internet giant Google, which

In 1986, Honda engineers set out to create a walking robot. The result was ASIMO, one of the world's most advanced humanoid robots. Its latest incarnation can push a cart, pour a drink and serve it to you on a tray.

could bring its expertise in mapping and voice, speech, image and face recognition to the robot. Indeed, one app they're working on allows a remote user to tell Ava to track down someone in the home and get them to answer the phone.

Yet some sensing technology still has stubborn problems. Speech recognition, for instance, is commonplace in commercial computing today, so you'd think it wouldn't be a problem for a robot butler to recognize verbal commands. But current speech-recognition technology is not foolproof. Siri has been a feature of Apple's iPhone since 2011, allowing you to access your music, contacts, send a text, schedule an appointment, or even search the web just by issuing a voice command. However, Siri has problems with accents, noisy rooms and understanding commands in different word orders. To avoid confusion, perhaps it would also be useful if our robot butler could track posture, hand gestures, facial expressions and gaze – cues we humans take for granted in our communication. Like speech recognition, we've been experimenting with face tracking and face-recognition technology for years. Results so far have been mixed, although there is some evidence that we can train a robot to infer human intentions from a face. Japanese researchers from the University of Tsukuba have been using a wireless headband that measures the electrical activity produced by the skeletal muscles. It can read smiles and frowns with 97 per cent accuracy and using this the team has got a robot to recognize if a person wants a ball handed to them or thrown based on their facial expression.

The array of sensing in development is impressive, but it doesn't solve the problem of how our robot is going to integrate all of this information at the same time – and make decisions based on the information. For that, it may require some form of intelligence.

AI ROBOT

In 1950, the British mathematician and scientist Alan Turing wondered whether machines could think. He explored this idea by posing a more precise question: "Are there imaginable digital computers which would do well in the imitation game?". This forms the basis of the Turing test: in a purely text conversation in which neither participant can see the other, can a human tell if his partner in conversation is a person or a machine? In 1966 the answer was "no". ELIZA, a program invented by the German scientist Joseph Weizenbaum, analyzed the words a human participant typed in, looking for keywords to which it could apply rules and generate a sentence in response. If no keywords were found, ELIZA had a pile of stock responses with which to stall, including just repeating an earlier comment. This was enough to convince several people that ELIZA was a person, but a problem with the Turing test is that humans can be fooled

The RoboCup's goal is to field a team of humanoid robots that, by 2050, can beat the reigning (human) men's world champions. For now it is an annual competition where roboticists compete to get their automatons to kick the ball, control the ball, signal their teammates and just keep from falling over.

pretty easily. Delivering plausible answers in a conversation doesn't require any intelligence at all – you just have to produce a reasonable semblance of human behaviour. We test this regularly when chatbot computer programs compete with human participants for the annual Loebner Prize. The aim is for the programs to convince a panel of (human) judges that they are the most "human". Nothing has ever won the top prize (for the first program whose responses are completely indistinguishable from a human's), but every year a few judges are fooled and a bronze medal is awarded for the "most convincing human-like conversation".

Chatbots are an example of "weak" AI (artificial intelligence) – machines that appear intelligent but aren't in the way that we are. Through intricate and meticulous programming, a large number of human behaviours can be replicated that could easily be mistaken for intelligence. But it's not the same as a "strong" AI – a machine that can replicate the features of the human brain in being conscious, self-aware and truly intelligent.

Intelligence is a difficult term – we don't know exactly what intelligence is or how it works (*see* QUESTION 5: WHAT IS CONSCIOUSNESS?) but this hasn't stopped researchers from trying to replicate it. Some are taking that literally, building brains neuron by neuron in "artificial neural networks" to see if intelligence emerges. These neural networks aim to mimic the way the brain learns: repeated stimulation of one neuron by another strengthens the physical connection between the two (and if not, the connection weakens). In a famous example, researchers at the University of California, San Diego, trained 920 artificial neurons to correctly conjugate the past tense of English verbs – no mean feat given that English has a large number of irregular verbs. It made some mistakes along the way ("go" to "goed" rather than "went", for instance), but these were similar to the mistakes children make. The idea of a machine learning as we do is one of the hallmarks of developmental robotics and it's an attractive one for our robot butler. While it's nice to get something that works straight away "out-of the box", this one-size-fits-all model isn't very adaptable. A robot that develops – learning to do things through its own trial and error or by being taught by a human (or perhaps another robot) – could pick up skills specific to its master's needs and continue to add new skills throughout its lifespan.

There are many other approaches to AI and it's only a matter of time before any of them, or a combination, delivers intelligence similar to our own. When that happens, we're going to have to deal with a range of ethical issues that sci-fi authors have been exploring for decades, not least how much we will trust a thinking machine. Karel Čapek's *R.U.R.* saw the robots rise up violently to overthrow and kill their human masters. One of the most famous ways to prevent such an outcome was proposed by another writer, Isaac Asimov, who, in his 1942 story "Runaround", introduced his "Three Laws of Robotics":

1. A robot may not injure a human being or, through inaction, allow a human being to come to harm.
2. A robot must obey the orders given to it by human beings, except where such orders would conflict with the First Law.
3. A robot must protect its own existence as long as such protection does not conflict with the First or Second Laws.

This seems a neat solution but "Runaround" shows that Asimov's Three Laws are not infallible. In the story, a scientist's failure to emphasize just how urgent a dangerous, but life-saving, task is throws the robot's AI into conflict. Stuck between its need to protect itself and its compulsion to obey its master, the robot runs around in circles, rambling incoherently. This highlights the potential pitfalls if we don't think through the intricacies of a robot's programming. In serving tea, for instance, how does a robot butler decide when to let go of the cup when handing it to its master? What if the master doesn't grip the cup correctly – does the robot let it fall or grip tighter? And say the master forgets what they've asked for – when does a robot decide that the transaction has failed? It's natural to ask how reliably we can trust robots, but robot cognition is difficult to create when we ourselves are not particularly reliable.

- - ➤

Robotic grippers, let alone humanoid hands, are a challenge in robotics, requiring sensing, dexterity and balanced gripping. Their development could also play a key role in human-robot communications.

INSIDE EXPERT:
HOW DOES THE PUBLIC PERCEPTION OF ROBOTS COMPARE TO REAL ROBOTICS?

"People have a pretty sophisticated idea of what robots are these days, but there's still a big expectation gap. There's an assumption that robots are much smarter than they really are. We scientists have over-promised, especially in AI, and failed to explain just how difficult some of the problems of intelligent robotics are. But there are a huge number of challenges, of which AI is just one. There's how to make robots soft and compliant so that they won't hurt you if they crash into you. We have robot hands approaching the number of degrees of freedom that a human hand has but they're still too heavy. Robots can pick up a cup, but that in itself isn't enough to solve the problem of how to safely hand it to a human. Consider the sensing problem. Even the most sensor-rich present-day robot falls far short of the simplest insect's sensorium. Animals have this amazing stuff called skin, which has millions of touch sensors embedded in it. We are only just beginning to address that particular problem in robotics.

The amazing thing is that a human can pick up a cup of coffee with her eyes closed, even if she's never seen that particular cup before. No robot could do that – it just doesn't have the sensing and cognitive capability to safely deal with the unknowns. The list of major problems facing robotics is still very long, and it's a holistic problem. You need the whole package to make a truly sophisticated robot – you can't just have great grippers, or great sensors, or a great AI – you need all of them, perfectly integrated; a deficit in any one will seriously impair the robot."

Alan Winfield, Professor of Electronic Engineering, University of the West of England, Bristol, UK

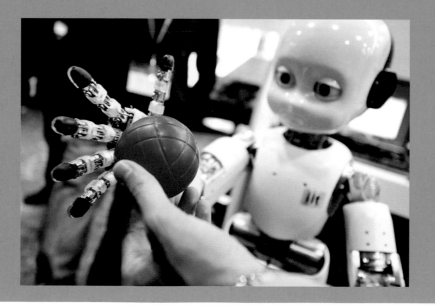

THE FUTURE IS NOW

Outside of cognition, there are still a great many problems to be solved before personal robots become practical. One is power – many industrial or commercial robots are either permanently wired to the mains or have to return regularly to a nearby docking station for charging. Another is price. It's one thing to make a prototype but it is quite another to turn this into a commercially viable product, something that can be mass-produced at a price most people would find reasonable. And given the optimization that goes into robots designed for a single purpose and for a single, closed, structured environment (which is most of robots), doing it for a single, multi-purpose, roaming robot butler would be even more expensive.

What seems more plausible is a staff of smaller, specialized robot helpers, similar to what Homer had imagined for Hephaestus over two millennia ago in the *Iliad*. From automated lawnmowers to driverless cars, many already exist and some are already in our homes. iRobot's Roomba robot vacuum cleaner, for instance, has been around since 2003. This flat, round robot is not particularly efficient, often covering the same spot several times, but it avoids obstacles and cleans every inch of exposed floor without its owner having to do anything besides empty the dust bag.

In fact, small teams of robots could prove pretty adaptable. Swarm robotics builds on the natural "swarm intelligence" seen in colony insects like termites. This approach is attractive because of its built-in redundancy, which gives the system a lot of robustness – even if one unit fails, there are still enough to get the job done. Having no central computer reduces the risk of something going catastrophically wrong in the whole system because of one central error. And it's also scalable – if you want something done faster, just add more robots. Swarm robotics could prove useful in clean-up operations, agriculture, environmental monitoring, exploration, and search and rescue. The potential is demonstrated in two robot systems developed by Marco Dorigo, co-director of IRIDIA, the artificial intelligence lab of the Université Libre de Bruxelles in Belgium. His s-bots were able to find and rescue a dummy child by banding together to grab the clothing, then self-organizing into chains to drag the dummy to safety. Dorigo's other robot, Swarmanoid, has three types of non-identical units: a foot-bot, a hand-bot and a flying "eye-in-the-sky" eye-bot, which can collectively locate and fetch a book from a shelf. Then there is Symbrion, in development by a collaboration of European universities. Like Swarmanoid, it has three non-identical units (backbone, active wheel and scout) but with a common docking interface and multiple docking points so each unit has the potential to dock with more than one other unit. Any number of these units can combine in different multicellular forms, devised on the fly to solve the

First unveiled in 2002, the Roomba robot vacuum cleaner has sold over 8 million units worldwide. It forgoes expensive sensors and fancy intelligent programming to keep the cost affordable for average consumers.

task at hand. Each unit's role changes from combination to combination, and they can act individually as well as collectively.

Swarm robotics is still in its infancy, but it's likely that a significant proportion of households worldwide will have at least one domestic robot of some kind within a decade. The International Federation of Robotics expects 15.6 million service robots to be sold for personal use between 2012 and 2015. If South Korea meets its 2013 goal, it may already have a robotic teaching assistant in every kindergarten. And in Japan – by 2007 estimates, home to 40 per cent of the world's robots – the government has publicly stated its intention to use autonomous service robots to care for its elderly by 2025. After 2,500 years, the musings of Homer and Aristotle might actually be realized.

Swarm robotics imagines not a single robot but a team of robots that are collectively able to perform any task. Robust and adaptable, this could be the future of robotics. - - ➔

14

HOW WILL WE BEAT BACTERIA?

There are few good things that come from little boys' bad habits of not cleaning up after themselves and examining their snot. Yet a grown man doing exactly that won the Nobel Prize. One September morning in 1928, Alexander Fleming returned to his basement laboratory at St Mary's Hospital in London, fresh from a summer holiday in Suffolk. Not being a particularly tidy man, he'd left his bacterial cultures in a corner while he was away. Mould had grown on one of the culture plates he'd accidentally left, and the mould had done something rather odd to the bacteria. Immediately around the *Penicillium* mould, colonies of *Staphylococcus* bacteria were either dead or absent entirely while those further away were fine. Fleming deduced that some chemical from the mould was preventing the bacteria from growing, even when diluted 800 times. He'd seen a similar thing years before: suffering from a cold, he'd cultured his own mucus and found that any bacteria near the mucus either had their growth inhibited or were destroyed. That led him to the discovery of the bacterial enzyme lysozyme, so he took the same approach with the *Penicillium* mould. With the help of his assistants, Fleming tested a crude mix of the bacteria-killing chemical from the "mould juice" and named it penicillin.

This was one of the landmarks in the history of medicine, forever changing the way we treat bacterial infections. By the 1940s, two scientists from the University of Oxford, Howard Florey and Ernst Chain, had successfully purified penicillin and proved its antibacterial properties in mice and humans. Penicillin's discovery, for which Fleming, Florey and Chain were awarded the 1945 Nobel Prize in physiology or medicine, is one of the great stories of science and serendipity. By the middle of the twentieth century had sparked an entire industry, producing medicines that would fight some of the deadliest diseases, including syphilis, diphtheria, gangrene, pneumonia, typhoid fever, meningitis and tuberculosis (TB). This, in turn, paved the way for other revolutionary advances in surgery, organ transplantation and cancer chemotherapy. It's no exaggeration to say that antibiotics are one of the miracles of modern medicine, saving untold lives and increasing human life expectancy by an average of eight years. The bad news is that the miracle might be at an end. The bacteria are fighting back, and it looks as if they're winning.

In the European Union alone, it's estimated that around 25,000 people die each year of infections caused by multi-drug resistant bacteria. Even in the best hospitals around the world, a high percentage of acquired infections are caused by highly resistant bacteria such as MRSA (methicillin-resistant *Staphylococcus aureas*) or *Clostridium difficile*. Around 440,000 new cases of multi-drug resistant TB emerge annually, killing at least 150,000 people worldwide. Worse, extensively drug-resistant TB has been reported in 64 countries and there are signs of what might be a totally drug-resistant strain of TB. And the worry is that, whether TB or other bacteria, there may already be new strains emerging that are untreatable by any existing antibiotic.

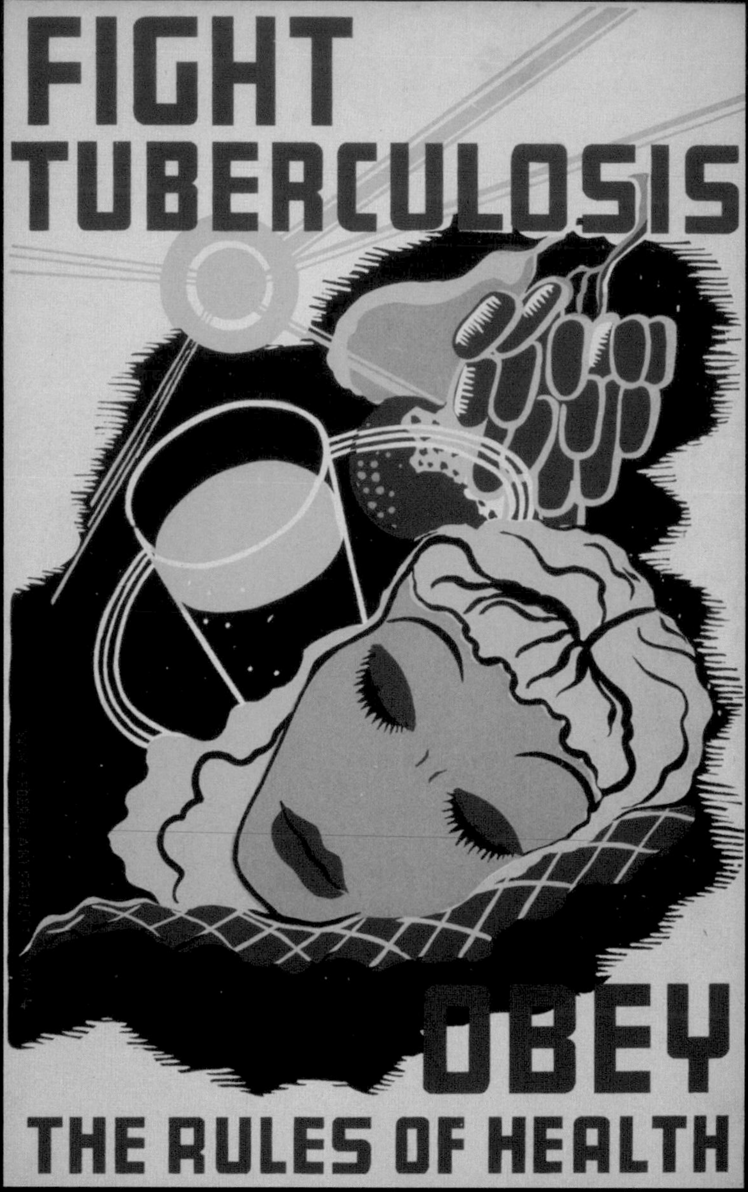

A US TB health campaign poster c.1936–41. This poster is promoting proper diet and adequate sleep and sunshine exposure to help prevent TB.

It's not just the direct effects of disease and death that are terrifying, but the knock-on effects that bacterial resistance has on medicine. Patients remain infectious for longer, which also dramatically raises the cost of health care, as does the need to use more expensive therapies if the front-line medicines fail. Such costs are catastrophic for countries whose health-care systems are already buckling. And just as antibiotics have allowed us

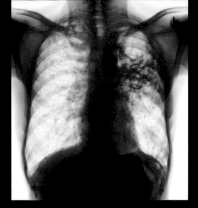

Coloured X-ray of the chest of a male patient with pulmonary TB. Caused by the bacterium *Mycobacterium tuberculosis,* which is spread by coughing and sneezing, the bacteria create primary tubercles – nodular lesions of dead tissue and bacteria – when inhaled into the lungs.

to make great gains in surgery and other areas, the loss of their protection jeopardizes all of them.

RESISTANCE IS NOT FUTILE

Bacteria that are resistant to an antibiotic are invulnerable to its effects, although the presence of such resistance should not come as a surprise – antibiotic resistance has been known to us almost as long as we've known about antibiotics. Fleming, for one, realized early in his experiments that bacteria developed resistance whenever too little penicillin was used or when it was used for too short a period of time. Antibiotics and antibiotic resistance go hand in hand – the creation of one inevitably produces the other. This makes sense, as bacteria make antibiotics themselves (although we're not entirely certain why – *see* INSIDE EXPERT WHY DO ANTIBIOTICS EXIST?). But the fact is that for every new antibiotic we discover, resistance already exists somewhere, even if it hasn't been switched on yet. Microbiologists speak of the "resistome", the reservoir of genes coding for resistance to any antibiotic we care to deploy. This reservoir can be found in the many harmless bacteria that are all around us, such as the *Streptomyces* bacteria that are found in soil and from which the vast majority of our antibiotics are made. All it takes for a drug-resistant form to develop is for a resistance gene to jump from a harmless soil bacterium to a disease-causing one.

How do bacteria become resistant? One way is a straightforward change in a bacterium's genetic make-up – mutation. The nature of genetics is such that small, random mutations occur every time a cell divides – all it takes is for one, or several, to effect some change that confers resistance. Take streptomycin, for instance. This antibiotic works by binding to a single target, a part of the cell called the ribosome that is involved in protein synthesis. The part streptomycin binds to is encoded by a single gene, so a single mutation could change that target and stop the antibiotic from working. Random mutation is part of evolution, and this is where the laws of natural selection work against us. The new antibiotic-resistant mutant bacteria are now the "fittest" and will survive, thrive and multiply after our antibiotic treatment kills off the non-resistant bacteria. Through the very act of using antibiotics, we cause bacteria to evolve resistance.

However, the main way bacteria become resistant to an antibiotic is by acquiring resistance genes rather than by mutation. Horizontal gene transfer refers to the swapping of genes between organisms without being passed on through reproduction (which passes genes on "vertically" through the generations). Bacteria can actively swap genes by using small, circular DNA molecules known as plasmids or by viruses called bacteriophages. Through these mediators, a bacterium can pick up any number of genes – including resistance genes – from any other bacteria around it.

BAD HABITS

Resistance may be a natural phenomenon but its rapid spread is of our own doing. For one thing, we've been using antibiotics when there really is no need. The antibiotics revolution was so profound as to seem like a panacea. Surveys in different countries have shown that patients turn up to their doctor expecting to be given antibiotics, irrespective of their actual illness. In part, this is due to the failure of doctors, scientists and health professionals to communicate exactly what antibiotics are, what they work on (bacteria) and what they do not (viruses, parasites and any disease that is not caused by bacteria). But alongside public misunderstanding, many doctors must also shoulder the blame for prescribing antibiotics without properly diagnosing the problem, issuing them "just in case" or to meet artificial targets for drug prescriptions, or, in some cases, just to keep the patient happy.

Alexander Fleming himself cautioned many times against the use of penicillin unless there was a good reason and, if it must be used, that it should be at the right dosage and for the right period of time. These last two points are vital (not just for penicillin, but for all antibiotics) to ensure the disease-causing bacteria are eliminated completely. If you don't kill every remaining vestige of the infective bacteria – perhaps because you don't complete the full course of antibiotics – the surviving bacteria acquire resistance, propagate and come back in force to make you, and everyone else you infect, pay for your mistake.

Some of the spread of antibiotic resistance is the result of globalization. Faster, cheaper travel means that people are moving about the globe at a spectacular pace. This is a boon for pathogenic bacteria who hitchhike on (and in) us to different towns and cities, regions and countries,

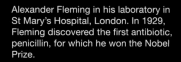

Alexander Fleming in his laboratory in St Mary's Hospital, London. In 1929, Fleming discovered the first antibiotic, penicillin, for which he won the Nobel Prize.

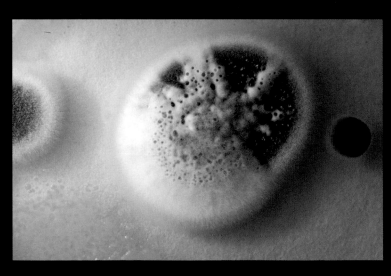

← - -

Colony of *Penicillium chrysogenum* fungus growing on agar in a petri dish. This fungus produces the antibiotic penicillin G, seen as small droplets on the surface.

KEY IDEA:
GOOD BACTERIA

When we think about bacteria, we immediately link them with disease. But in fact there are plenty of good bacteria that help us out – a trillion of them live symbiotically in our gut, aiding digestion and immunity. Furthermore, scientists think that knowing more about these good bacteria might open up a new front in the war against antibiotic resistance.

Unfortunately, most antibiotic therapies are indiscriminate: they destroy the good as well as the bad bacteria. One study conducted at the University of València in Spain found that it took four weeks for the gut bacteria to re-establish themselves after antibiotic treatment and certain good bacteria did not reappear at all. One way to restore the balance might be through probiotics. There's a big commercial market for these, though if you're healthy your gut bacteria are fine so there's no need to buy probiotic drinks or yoghurt. But if you've had to undergo antibiotic treatment, particularly a non-specific one, probiotics can be helpful in repopulating your gut with the many good bacteria the antibiotics would have taken out as collateral damage.

A more unorthodox approach is faecal transplants. As gross as this sounds, it's actually very viable. Around 60 per cent of your faeces is made up of bacteria from your gut and transplanting the good bacteria from a healthy patient to a sick one could combat malign bacterial infections. Faecal transplants have been used to treat hundreds of patients, more than 90 per cent of whom have recovered. In fact, one 2012 trial was so successful that it was halted midway – the faecal transplants were found to be three to four times more effective than an antibiotic and cured 15 out of 16 patients infected with *Clostridium difficile*. (Of the patients treated with an antibiotic, seven out of 16 were cured; the rest were treated afterwards with faecal transplants, which cured them.)

landing in new areas where they can infect vulnerable humans and maybe swap resistance genes with the local bacteria. This is, to some extent, unavoidable in the face of progress, but other mistakes are less forgivable. Farmers have been adding antibiotics to animal feed since the 1940s, not only to keep their animals disease-free but also as "growth promoters". In the US, a staggering 80 per cent of all antibiotics sold are estimated to go to the farming industry. With hindsight, such brazen use of antibiotics seems incredibly foolish and, thankfully, we are trying to rectify the matter: the European Union banned the practice in 2006 while the US introduced new (voluntary) regulations in 2012. We're finally getting smarter at using antibiotics, but this is where we get another bit of bad news: the pipeline of new antibiotics is running out. In fact, it started sputtering a while ago.

DRY PIPELINE

Selman Waksman is a name synonymous with antibiotic discovery. The Ukrainian scientist discovered the vast majority of our existing antibiotics by painstakingly testing the compounds that soil bacteria, such as the *Streptomyces* species, produced in competition with each other. The 20 compounds he and his team discovered include streptomycin, the first antibiotic found after penicillin and the first drug that could treat TB, for which Waksman received the Nobel Prize in physiology or medicine in 1952. These compounds were the foundation of the golden age of antibiotic discovery in the 1940s and 1950s.

Unfortunately for us, the golden age quickly lost its shine. Only a handful of new antibiotics have entered the market in the last 50 years, so we haven't got many new weapons in reserve to tackle resistant bacteria. The pipeline of new antibiotics is dry and there's little incentive to get it flowing again. Part of the reason is that discovering new antibiotics is very hard. In the golden age, people sought out bacteria from different environments, such as beaches, river sediments and the craters of volcanoes. They found such an abundance of new antibiotics that some scientists thought we'd found all there was to discover. Many drug companies shut down their antibiotic discovery programmes, which makes financial sense because antibiotics are not a long-term money-maker. Resistance means that the window for an antibiotic's effectiveness is very narrow – within two years of entering the market, an antibiotic will already have lost its edge. Even if it's a highly effective new one, it will probably be saved for emergency use when no other antibiotic will work, so the number a drug company can sell is limited. And even when an antibiotic is prescribed, patients take it for only a limited time. Moreover, most antibiotics retail for low prices and the patents granted on

Selman Waksman, the discoverer of many antibiotics in the "golden age" of the 1940s and 1950s. Waksman was awarded the Nobel Prize in 1952 for his discovery of streptomycin, the first antibiotic treatment for TB.

huge amount of resources, and it can take at least a decade to get a drug to market. We need the pharmaceutical industry but it's difficult for any company to justify the risk of investment in new antibiotics, particularly compared to medicines for chronic conditions like heart disease, which generally have to be taken for long periods and so produce much greater financial return.

NEW DRUGS

The development of new antibiotics may have dried up in recent years, but that doesn't mean there aren't new ones to be discovered. We have so far just harvested the "low hanging fruit". Now we realize that all we'd discovered was the first antibiotic made by a particular strain of bacteria, often the most prevalent one or the one that happens to appear under

INSIDE EXPERT:
WHY DO ANTIBIOTICS EXIST?

"We humans tend to think of antibiotics in the context of medicines to fight bacterial infection, so the traditional thinking is of antibiotics as chemical weapons, helping bacteria fight off the competition for resources such as water, oxygen and food. And soil bacteria have a lot of competition – there could be around a trillion microbes in a gram [about 0.04 ounces] of soil. But why would they make so many different antibiotics if all they want to do is kill their neighbours? Streptomyces bacteria can make, on average, around 12 different antibiotics with very diverse structures and modes of action. Why go to all that trouble when something cruder would work? Either they're using multi-drug therapy (in which case they're much cleverer than us) or these things act in different ways that we don't understand.

Another school of thought is that antibiotics may actually be chemical signals allowing bacteria to communicate with each other. We use antibiotics at artificially high concentrations in the clinic. The concentrations bacteria produce in the natural environment are a lot lower and have very subtle effects on gene expression and protein production in the bacteria. So antibiotics could actually be signalling molecules to other bacteria, particularly related ones.

My feeling is that antibiotics probably act as both: chemical weapons against bacteria that aren't resistant and signalling molecules to related bacteria, telling them to act a certain way or even produce the same antibiotic to kill off other microbes."

Dr Matthew Hutchings,
Senior Lecturer,
University of East Anglia, UK

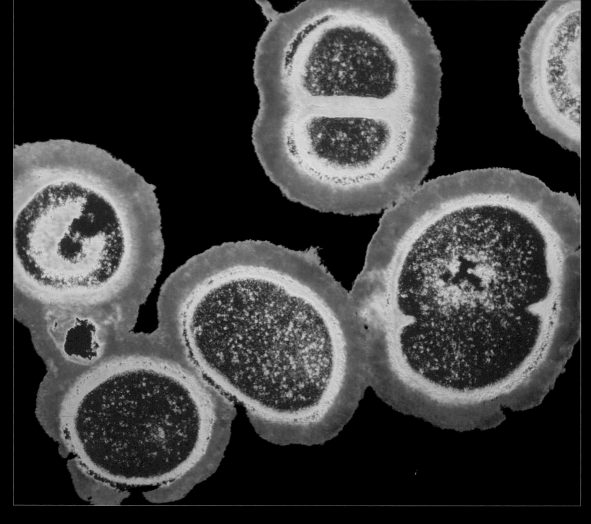

laboratory conditions. The advent of DNA sequencing means that we now have powerful tools to discover just how many more there are. Genome studies show us that *Streptomyces coelicolor*, for example, can make more than 20 potential antimicrobial substances but makes only four under laboratory conditions. The chemical pathways that make the others are turned on in the environment, but we don't know what the switches are and so can't turn them on in laboratory conditions. There's now a research effort to activate these "silent pathways" in the known strains of *Streptomyces* – and there are 500 known species kept in laboratory freezers across the world. This represents a huge number of potential new antibiotics and is the focus of many small start-up companies spun out of university research groups. If these new companies discover any promising

MRSA (Methicillin-resistant *Staphylococcus aureus*) bacteria dividing. MRSA is common in hospitals, infecting the wounds of patients. Worryingly, strains of MRSA are resistant to most antibiotics.

new drugs, it may well become viable for a pharmaceutical company to buy them, without having to put in the initial research and development themselves. There are also efforts to lobby governments to relax clinical trial requirements and extend the length of patents for new antibiotics, thereby making it cheaper, easier and more attractive for companies to develop them.

It may also be possible to get more mileage out of old antibiotics by finding new ways to activate them. Canadian biochemist Gerry Wright's group at McMaster University in Ontario tested different combinations of (non-antibiotic) drug compounds in tandem with minocycline, an antibiotic long abandoned due to resistance. Remarkably, they found 69 compounds, never before used to treat bacterial infections, that helped minocycline regain its effectiveness. One of the compounds they've tested is the anti-diarrhoea drug loperamide, which helps minocycline significantly inhibit the growth of *Pseudomonas aeruginosa* (which can cause a variety of diseases in humans, particularly in people with low immunity) and *Escherichia coli* (a common bacteria, some strains of which can cause food poisoning and other illnesses) by weakening the bacteria enough so that the antibiotic can work.

Others are looking at the imporant relationship our body has with certain bacteria (*see* KEY IDEA: GOOD BACTERIA). And people are still looking for new antibiotics in weird and wonderful places. Stories about potential antibiotics found in crocodiles, cockroach brains and panda blood make good headlines, but their practical use is unknown. A more realistic avenue is looking for symbiosis, where bacteria have co-evolved to live with a plant or animal host in a mutually beneficial relationship, with the bacteria providing protection against infection from other microbes. These might be found in environmental niches unexplored in the golden age, such as the deep sea (*see* QUESTION 16: WHAT'S AT THE BOTTOM OF THE OCEAN?), where bacteria might live in primitive marine animals like sponges, cone snails and coral reefs. Even on land there are soil bacteria that prevent plants from getting infections. These relationships have been forming for tens of millions of years, so it's possible that they have evolved new biochemical pathways which we haven't discovered. And they may use antibiotics in a clever way too, perhaps switching bacteria regularly and/or using different cocktails of antibiotics as multi-drug therapies. This is something we're trying ourselves – any bacterium that has resistance to one antibiotic probably won't have resistance to another, and gaining resistance to a whole cocktail of antibiotics at once is rather unlikely.

Yet the truth is we will never "beat" bacteria. One of the earliest forms of life on Earth, they were around over three billion years before us and will likely live on long after the human race is gone. Our co-existence with them is an arms race that we will never win in the long run, but we can certainly keep abreast if we get smarter in our use of the best weapon we have: antibiotics.

15

WILL WE EVER CURE CANCER?

There's no easy way to say this: you're going to die and the chances are you're going to die from cancer. What's more, cancer will never be eradicated. It's been around so long that fossil evidence suggests even the dinosaurs may have had cancer, while the earliest human records date back to an Egyptian papyrus, written in around 3000 BCE, describing a bulging tumour of the breast with no known treatment. In 2008, the disease accounted for 7.6 million deaths worldwide – 13 per cent of all deaths that year – and sometime soon cancer will overtake heart disease to become the biggest killer globally.

Cancer's terrifying ubiquity comes from the fact that the potential to develop the disease is hardwired into each and every one of us. Almost any cell can turn into a cancer cell. When this happens, its machinery drastically malfunctions. For instance, some cells might pump out hormones, which may lead to the release of huge quantities of steroids. Others may cause low blood sugar, or they might dump calcium into the blood, with toxic consequences for the heart and nervous system. As these rogue cells multiply uncontrollably, they spread to other tissues around the body, putting pressure on organs and further disrupting body functions.

The thing about cancer is that it's extremely complicated. It's not a single disease – cancer is a loose term for hundreds of different diseases, all with subtle differences in their causes and effects; breast cancer alone can be categorized into at least 10 distinct types of tumour. There isn't a single cause of cancer, either. It's the result of a combination of a person's inherited genetic make-up, genetic faults picked up over the course of a lifetime and a mixture of other risk factors. These include smoking, obesity, diet, alcohol, level of physical activity, chemicals, bacteria, viruses, too many hours spent in the sun or on a sunbed, and many more. On top of these individual factors, cancer seems to affect people differently depending on their age, sex and where they live. Whatever the causes, the result is that the cell's genetic material (its DNA) is changed (mutated), causing its machinery to run amok, turning on genes that encourage a cell to multiply and turning off genes that would normally stop runaway growth. Repair mechanisms fail and, left unfixed, the damage accumulates. And the longer we live, the more we're exposed to factors that might cause mutations and the less efficient our cells become at fixing the damage. This is why cancer is usually associated with ageing (*see* CHAPTER 18: CAN WE LIVE FOREVER?). So although some people are unfortunately born with mutations that predispose them to certain cancers, there is no one mutation that triggers a cancer; several have to build up for a tumour to start growing, and more for it to spread.

Finding all these mutations is a colossal task, so cancer geneticists decided to form a "dream team". Working together on the Cancer Genome Project, they've found that just over 1 per cent of all human genes could, if mutated, potentially cause cancer. One mutation, known as the BRAF mutation, is found in some 7 per cent of all human cancers, while mutations in the BRCA genes greatly increase the risk of developing breast, ovarian or prostate cancers.

The dream team has since gone global, with an international cancer consortium sequencing in full the genomes (DNA codes) of 25,000 patients, who between them have 50 different types of cancer. Things got off to a good start in 2009, when scientists at the Wellcome Trust Sanger Institute, near Cambridge in the UK, published the DNA sequences for lung and melanoma (skin) cancers from two patients. They catalogued 90 per cent of the mutations in the cancerous tissue, finding more than 33,000 in the melanoma cell and 23,000 in the lung cancer cell. The effects of tobacco smoke were all too clear: most of the lung cancer mutations were caused by chemicals in cigarette smoke, equating to one mutation for every 15 cigarettes smoked.

There is now a deluge of data coming out of genetic studies, and this brings the challenge of discovering its significance for cancer. We need to identify the mutations that are crucial to a cancer cell's survival – those that are "drivers" of its growth and those that are merely "passengers" with no effect. One study of 100 breast cancer patients found 40 different driver mutations in 73 different combinations, with each patient having between one and six mutations. However, one particular mutation was found in only 28 of the patients, which illustrates the complexity of the problem: mutations driving cancer

The Edwin Smith Papyrus, dated 3000 BCE and originating in ancient Egypt, contains the first recorded mention of cancer: a bulging tumour of the breast with no treatment.

Scientists at Cancer Research UK's Cambridge Institute in Cambridge, UK.

in some patients are not present in others. The BRAF and BRCA mutations are two of the more common cancer mutations but even they aren't found in every person with skin or breast cancer. This has implications for drug treatment: drugs that target a particular mutation will not help those who do not have that mutation. Trastuzumab (sold under the brand name Herceptin) is one of the best cancer drugs we have, but unless you have the HER2 mutation, it won't do much good. Similarly, gefitinib (sold as Iressa) is effective against various cancers but works only in people with the EGFR mutation. The challenge is not only to identify mutations and find ways of targeting them, but also to find efficient ways to screen patients so that we can treat the right people with the right drugs.

EVOLUTION

It's 1837 and Charles Darwin is pondering how different species fit together. Opening up his notebook, he sketches a multi-branched tree, the points of which represent different species and the branches showing which are related. The sketch was published in Darwin's famous book *On the Origin of Species* in 1859 and the "tree of life" became an iconic metaphor for his theory of evolution. Fast-forward to 2010 and another Charles, Swanton of the London Research Institute, is tracing the origins of one particular patient's cancer cells back to their key driver mutations. He's creating a map of how the mutations may have changed over time. And it looks just like the tree of life.

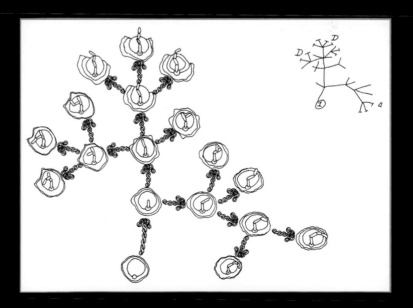

One of the reasons cancer is so hard to treat is that it is evasive on two levels: it can develop resistance to any drug we care to use, and it is constantly changing its genetic make-up. It is never the same for long and is constantly evolving. One of the most important realizations of recent years is that, even within the same tumour in the same person, one cancer cell's genetic make-up can be quite different from another's. In 2012, Swanton's team found that for one kidney tumour from a single patient around two-thirds of the mutations differed from each other. And when they looked at cancer cells that had metastasized (spread to other organs), the mutations were different again. So even when we detect a drug target in a tumour, the same drug may not work by the time the cancer has spread to other parts of the body. The key to getting around this might be in that tree of life diagram. Swanton says it underscores the importance of identifying and targeting the common mutations found in the "trunk" of the tree – as opposed to those found in the "branches". And we've got to be thorough; each time we kill some cancer cells but fail to destroy them all, those left will grow stronger, regroup and re-emerge.

What causes a tumour to spread? Different cancers spread in different ways to different organs. The lungs and liver are often first to fall; like port towns vulnerable to invading warships, their large blood supply puts them right in the firing line. And while most travelling cancer cells do not survive long enough to successfully colonize new regions of the body, it happens often enough to be a major problem. Any metastasizing cancer cell needs a supply line to establish its colony, so it does what many invaders do: pillage. Once the cell has settled, it starts producing chemicals that encourage the growth of blood vessels to bring in supplies from surrounding cells. Scientists hope that, by targeting the factors involved in blood vessel growth, they might be able to strand them on foreign soil, unable to replenish their supplies and vulnerable to a drug attack.

Surgery remains the most effective treatment for cancer, but removing just the tumour, while leaving other tissue unharmed, remains a challenge, particularly in the brain.

– – ➔

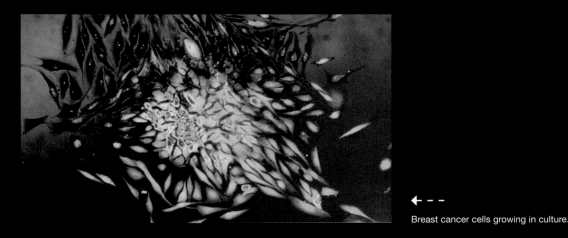

← – –
Breast cancer cells growing in culture.

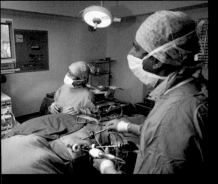

WHAT CAN WE DO ABOUT CANCER?

Attempts to cure cancer have a long history. The ancient Egyptians tried cautery (burning), knives, salts and arsenic paste. The ancient Sumerians, Chinese, Indians, Persians and Hebrews tried herbal remedies, including teas, fruit juices, figs and boiled cabbage. By around the fourth century BCE, in the time of Hippocrates – the father of medicine who named cancer for its likeness to a crab – the Greeks had progressed to surgery. Today, this is still one of the best and most straightforward ways to treat cancer, although its success depends on early diagnosis. The key to successful surgical treatment is to remove no more tissue than is absolutely

INSIDE EXPERT:
WHAT ARE THE BIGGEST CHALLENGES FACING CANCER RESEARCHERS?

"We've made huge progress in understanding and treating cancer over the past half a century – long-term survival has doubled in the UK over the past 40 years. But in some types of cancer progress has been poor and we still have a very long way to go. As we learn more about the molecular nuts and bolts of cancer, it sometimes feels that the picture is getting ever more complicated – for example, how do we sift out the important mutations from the mass of [genetic] sequencing data? And what about the role of other types of information, such as epigenetic marks [heritable factors that are not part of DNA] that help to control gene activity? It's a huge challenge to turn this explosion of data and knowledge into more effective treatments for patients.

As well as pushing forward with research, we also need to think smarter and do more with what we already have. Most cancer treatments are not given in isolation, but in combination with each other. Testing new combinations of existing or shelved drugs could bring big benefits, especially for rarer cancers. And simply increasing access to the latest treatments – be it drugs, radiotherapy or surgery – would help a lot. More effective tobacco control

measures, such as standardized packaging, and raising awareness of the warning signs of cancer among the public and health professionals would also make a difference.

But the biggest challenge remains trying to understand the 'rogue organism' that is cancer. We need to think about cancers not in reductionist terms of individual rogue cells growing in a laboratory dish, but as organisms that are constantly evolving within us. What makes their DNA unstable and prone to mutation, and how do they adapt to pressures such as drug treatment or radiotherapy? Why do cancers start spreading, and why do they colonize some places and not others? How do they subvert the healthy cells around them – the so-called microenvironment – and evade the immune system? Learning the language cancer uses to talk to our body will offer new clues for future treatment.

There are still many unknowns about cancer, but the good news is these are increasingly 'known unknowns'."

Dr Kat Arney,
Science Information Manager,
Cancer Research UK

necessary, particularly in the brain, without leaving behind any cancerous cells. To this end, scientists are developing increasingly sophisticated imaging and detection tools to more accurately define the boundaries of tumours. One possible future development is an "intelligent knife". Common surgical practice uses a heated blade to seal blood vessels and so reduce blood loss. It may be possible to analyze chemicals in the smoke given off by the knife to determine the make-up of the tissue it is cutting – whether it is cancerous tissue or normal, healthy tissue.

Surgery remains the bedrock of cancer treatment, but another key method is radiotherapy. This uses radiation to kill cancer cells, and cures around 4 in 10 cancer patients. The trouble is that radiation is harmful to normal cells too, so there's a fine line between destroying the cancer and harming the patient. Pioneering techniques mean that doctors can shape radiation beams so that they closely match the shape of a tumour, use high-tech imaging to target clumps of cancer cells as they move within the body or use multiple fine beams of radiation to destroy tumours with sniper-like precision.

Some scientists are also looking at ways of enhancing our body's natural defences to better recognize and attack cancer cells. One approach is to boost immune cells called T-cells, which (among other functions) find, bind to and destroy cancer cells. Other scientists are developing cancer-seeking "homing missiles" – radioactive materials attached to antibodies that bind directly to cancer cells. Unfortunately, most patients don't respond to these immunotherapies. Few offer more than a one in 10 success rate, and many come with serious side effects.

Thankfully, there have been great successes in chemotherapy with drugs like tamoxifen, a breast cancer drug targeting the female sex hormone oestrogen, which is known to promote cell division and therefore tumour growth. Not only is this drug highly effective at killing cancer cells, it also has a preventive effect for women diagnosed with a certain type of breast cancer. For these women, taking tamoxifen daily for five years reduces the chance of dying from the cancer by a third, with the effect continuing even after a patient stops taking it. Taking it for 10 years after diagnosis reduces their risk even further.

As a result of the insights from genetic research, new drugs are being developed that are targeted at ever more specific mutations and their effects. The discovery of the BRAF mutation, for example, has given hope that BRAF inhibitor drugs, which target the faulty protein produced (an enzyme that helps cells multiply), could halt tumour growth in malignant melanoma, colorectal, ovarian and thyroid cancers. Yet the results have been mixed: some BRAF inhibitors have reduced tumours within weeks of treatment, only to see them return stronger and more resistant a few months later.

Perhaps the future for cancer treatment lies in delivery. A nanoparticle form of the anticancer drug doxorubicin (known as liposomal

Researchers at the Welsh School of Pharmacy at Cardiff University are among those looking to accurately target cancer cells with nano-scale delivery mechanisms.

- - - →

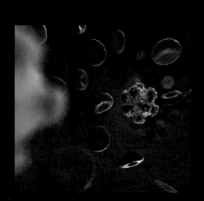

↑ A depiction of Human Papilloma Virus in the bloodstream. Almost all cases of cervical cancer – which kills around 250,000 people each year – are caused by HPV infection.

doxorubicin) delivers the drug directly to the tumour area. This method helps the drug evade the body's immune system, which might otherwise destroy the drug, while crucially avoiding the heart, where doxorubcin can cause severe side effects. The beauty of nanotechnology is its potential for custom engineering. One group of scientists at Harvard University is using DNA to make drug-filled "lock-boxes" that release their contents only when they meet the right "key", such as a molecule on the surface of a cancer cell. Requiring two or more keys would further increase its safety – like a nuclear missile that requires two verification keys before it can be fired.

Many of these technologies are still in the very early stages of testing, but the number of possible therapies is encouraging. We will never cure humanity of cancer, but reducing its burden is achievable. One in six of the world's total number of cancer cases is due to infections such as human papilloma virus, hepatitis B and C, and *Helicobacter pylori*, which can cause cervical, liver and stomach cancers, respectively. Many could be prevented through cheap vaccines and basic screening, particularly in poorer countries, which account for around 70 per cent of all cancer deaths worldwide. And here's the thing: between a third and a half of all cancers – 2.4 to 3.7 million a year – are actually preventable. More than 30 per cent could be prevented by quitting smoking, drinking alcohol moderately at most, eating a healthy diet, keeping physically active, avoiding the use of sunbeds and limiting exposure to direct sunlight. Cancer science has come a long way, but its most important findings are that a healthy lifestyle and awareness of the early signs (followed by prompt treatment) can make a world of difference.

WHAT'S AT THE BOTTOM OF THE OCEAN?

If the Earth were a giant beach ball, the oceans would barely moisten its surface. The deepest spot in the ocean – Challenger Deep in the Mariana Trench, east of the Philippines – is just 11 kilometres (7 miles) below the waves. That doesn't seem very far; it's the distance you might drive on a round trip to the supermarket or cover during an afternoon's easy ramble in the countryside. But going that far down – underwater – is not quite as straightforward as driving to the supermarket or rambling.

For one thing, the human body is not designed to deal with the underwater environment. Consider what happens to the free-diver who dives a few metres down. As soon as he hits the cold water, his heart shifts to a lower gear and his pulse rate plummets. Blood rushes away from his skin to his vital organs. At 5 metres (about 16 feet) down, still holding his breath, his brain might spontaneously turn itself off, causing a blackout. His lungs and spleen shrink as the immense weight of the water presses down on his body. Descending to 20 metres (65 feet) and deeper, the pressure continues to build, forcing any air that was keeping him afloat out of his wetsuit and forcing him downwards at speed as he loses buoyancy. Some of the most disciplined free-divers have reached depths of 100 metres (330 feet) and beyond, but even wearing full scuba gear won't enable you to go anywhere near as deep as a kilometre (about 3,280 feet).

What this means, of course, is that travelling those trivial 11 kilometres is fraught with all sorts of difficulty and danger. Getting a person to Challenger Deep is a challenge indeed, one requiring an impenetrable vessel capable of withstanding pressures of up to 1,100 atmospheres – equivalent to about 1.2 tonnes pressing on every square centimetre (8 tonnes on every square inch) of its hull. Since the two-man crew of a metal bathyscaphe called the *Trieste* reached the bottom there in 1960, only one manned vehicle has followed: the film-maker James Cameron, famous for *Titanic* and *Avatar*, made humankind's return in 2012. Fewer people have visited Challenger Deep than have visited the Moon and, unlike the Moon, no one has ever set foot on it.

The vast majority of the very deep ocean remains unexplored, yet tantalizingly within reach. Embarking on a Census of Marine Life in 2000, scientists estimated that only 5 per cent of the ocean had been searched for life. A decade of discovery under the Census yielded more than 20,000 new species, but what has been glimpsed so far of life on the ocean floor is just a tiny snapshot of a world that lies beyond our imagining. Only in the last few decades have scientists started to take a more eager interest in what might lurk at the bottom of the deep ocean, believing until recently that few species could survive in such an inhospitable environment – the cold, the dark, the crushing weight of all that water. However, while we know that the same diving response that prevents humans from reaching the deepest depths also prevents other mammals, creatures less like ourselves seem quite at home on the seafloor.

During their brief, half-hour stay at Challenger Deep back in 1960, US Navy Lieutenant Don Walsh and Swiss scientist Jacques Piccard spied jellyfish and what they took to be a layer of algae through the acrylic window of their creaking bathyscaphe. The fish that darted away at their disturbance were a more surprising discovery, since, like us, fish have a skeleton. At depth and high pressure, spineless creatures like jellyfish fare well because their flabby bodies are flexible to compression, but bones are less so. The deep-sea anglerfish, inhabiting sparsely populated regions some 3 kilometres (about 2 miles) or so below the surface, has evolved a soft skeleton to deal with this problem. It also lacks the swim bladder that fish which inhabit shallower water use as buoys – under such great pressure this air sac would collapse.

TEETH CHATTERINGLY COLD... AND SMOKING HOT

Since our rigid bones exclude us from a more intimate exploration of the very deep sea, those wanting to study its inhabitants have sent unmanned, remotely operated vehicles (ROVs) in their place, although few have plunged anywhere near as deep as the *Trieste*. In shallower waters, around 5 kilometres (about 3 miles), manned dives are easier, and the US Navy's *Alvin* research vessel has to date made nearly 5,000 dives, altogether carrying around 14,000 people up and down over a period of almost half a century. In the late 1970s, *Alvin* began ferrying scientists to hydrothermal vents known as "black smokers" – underwater chimneys that release large plumes of boiling black liquid (*see* BIG DISCOVERY: THERMAL SPRINGS SUPPORTING LIFE ON THE OCEAN FLOOR). Just as on dry land, there is volcanic activity below the Earth's crust at the bottom of the ocean. The vents form from minerals in seawater that has mixed with sulfide materials after seeping into cracks in the ocean floor; the black fluid that splurges out is at a near-volcanic temperature, heated by underlying molten rock.

The heat and peculiar chemistry of black smokers make for an unlikely living environment, but as researchers aboard *Alvin* and data from other vent-frequenting vessels are revealing, these smokers are actually hotbeds for life. Bit by bit, scientists are piecing together a picture of the unique ecosystems that exist at hydrothermal vents. The deepest, hottest vents we know of – discovered recently by a team of British scientists – are located 5 kilometres (about 3 miles) down in the Cayman Trough, a cleft in the floor of the Caribbean Sea. The same team embarked on a ROV tour of Southern Ocean vents at the East Scotia Ridge in January 2010 and returned with reports of species never seen before, including anemones, starfish and a pasty-looking octopus. Elsewhere, vents appear to be densely populated by communities of crabs, worms, mussels and shrimps.

Don Walsh and Jacques Piccard, the first men to journey to the bottom of the Mariana Trench, sit tight in the bathyscaphe. The pair spent most of the nine-hour dive perched on stools, telephoning up to the surface periodically and staying for just 20 minutes on the sea bottom.

We are learning that evolution can circumvent even the harshest conditions – for vent-dwellers and other species that have had millions of years to adapt to their surroundings, life in the deep sea poses fewer challenges than we might expect. Many deep-sea species, for instance, have adopted ingenious ways to deal with the lack of light. Below 200 metres (about 650 feet), sunlight barely pierces the gloom; at a kilometre (about 3,300 feet) down, light no longer penetrates at all and it is pitch black. This absence of light leads to a couple of problems. First is the problem of obtaining energy. All land-dwelling species get their energy from the Sun. Even carnivores, which obtain their energy by eating other animals, are dependent on light, because they are only ever a few links away in the food chain from their herbivore prey, which in turn feed on plants. Sitting right at the bottom of the food chain, plants are responsible for making the whole thing work: they are the photosynthesizers, harnessing the energy in sunlight and using it to make sugar from carbon dioxide and water so they can survive and grow. Photosynthesizers are not exclusive to

Invented by Jacques Piccard's father Auguste Piccard, a balloonist and physicist, the bathyscaphe was a research vessel designed for deep sea dives. The *Trieste* made dives off the Gulf of Naples and San Diego before Project "Nekton" – the mission that took her to the very bottom of the ocean.

the land. Light-harvesting, blue-green algae are widespread in marine environments, but in the lightless abyss of the deep sea the life forms at the bottom of the food chain must find another way to survive and grow – they must perform some other chemical trickery. The second problem is the more obvious one: vision. Without sunlight, how does a predator on the hunt for a tasty morsel ever find one, and how does a tasty morsel know to swim away?

Some progress toward solving the first problem was made when scientists realized that life around hydrothermal vents is based on tiny microbes that do something similar to photosynthesis, except without the light. They can turn chemicals straight into sugar – chemosynthesis, not photosynthesis. The rich chemistry of the vent environment provides the perfect energy source and the microbes thrive on it, blooming in their billions to form a thick, snowy layer that smothers the seabed around the vent. All other life at the vent depends on this chemosynthesizing mass: the microbes are fed on by shrimp, which are fed on in turn by fish and other, more exotic life forms. Although scientists are learning more about the locations and biology of vent communities, what is becoming clear is that they provide a lifeline for species stranded at the bottom of the ocean. Dotted along volcanic ridges straddling the world's ocean floor, they are island hubs for life and activity in an otherwise barren landscape – places where predators can find a regular supply of food rather than having to wait days for an unwitting fish to swim past or for a rotting corpse to descend from the upper ocean.

As well as getting their energy from chemicals, some deep-sea species generate light from chemicals, thus solving the second problem. Fuelled by molecules known as luciferins, bioluminescence is light produced entirely by biological organisms, a phenomenon remarked upon by the great thinkers of ancient civilizations: Aristotle recognized the "cold light" of bioluminescent fish millennia before the crew aboard *Alvin* did. Bioluminescent species exist from the top to the bottom of the ocean, helping predators to seek out prey and helping prey to avoid predators. The soft-skeletoned anglerfish has a bioluminescent, bacteria-filled "lamp" on its head, dangling from a flexible appendage between its eyes. Its function is to lure prey. While the anglerfish stays motionless on the seafloor, some hapless sea creature is attracted to the lure, thinking it has found food, but seconds later it finds itself being attacked and eaten by the anglerfish. Similarly, the cookie-cutter shark produces a bioluminescent pattern below its mouth that seems to resemble the silhouette of prey – potentially as a lure for smaller fish. Meanwhile, species that are unable to produce their own light have evolved the ability to collect others'. The bizarre barreleye fish has a transparent head that allows light to enter through the top as well as the front and sides. Until recently, scientists thought the fish's eyes

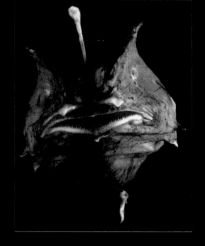

The deep sea angler fish has a soft skeleton that can flex under the pressure of the deep sea and a "lure" on the front of its head, filled with light-emitting bacteria, that it uses to attract prey.

Bioluminescent sea creatures use a molecule called luciferin, which combines with oxygen to produce oxyluciferin and light. More than half of the world's jellyfish are thought to be bioluminescent.

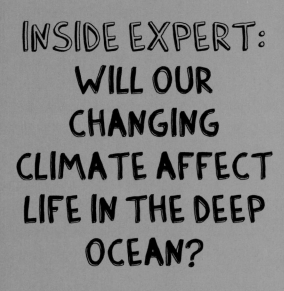

INSIDE EXPERT: WILL OUR CHANGING CLIMATE AFFECT LIFE IN THE DEEP OCEAN?

"I think the deep sea might serve as some sort of buffer – animals living in shallow waters will feel the effects of climate change earlier on. However, the first results from experiments in the North Atlantic off Svalbard have documented an increase in temperature in the deep water. It is just a fraction of a degree over a decade, but if this accumulates over a century, then it will impact on deep-sea organisms. The problem is that it is difficult to investigate the effects of potential changes, as it is almost impossible to obtain live animals from the deep sea. However, recent projects like the Census of Marine Life have really increased our knowledge of life in the deep sea tremendously, and the numbers of different species we have found just in small areas have exceeded our expectations. This knowledge is important, especially when we think more generally about impacts that mankind has on the deep sea, like deep-sea mining. There are so many animals and we don't really know anything about their biology – what they feed on, their adaptations or their roles in food webs. They have a role on the planet and nobody knows what will happen when we really start exploiting the deep sea and a lot of these species vanish."

Professor Angelika Brandt,
deep-sea biologist,
University of Hamburg, Germany

were fixed in a forward stare, but in 2008 scientists at the Monterey Bay Aquarium Research Institute in California proved that its eyes can swivel upward to keep a look out for prey approaching from above.

MARINE MIRACLES

Although the ocean floor remains largely unexplored, scientists have already found some astonishing things in the deep sea. But asking what's at the bottom of the ocean is like asking what's down the back of your sofa – there really is no telling unless you look, and is it worth the bother anyway? For some, it's about the spirit of human endeavour. Don Walsh and Jacques Piccard didn't journey to the bottom of the sea just to catch a glimpse of mysterious fish, and James Cameron was motivated by something more powerful even than money. In December 2012, Cameron told a crowd at a meeting of the American Geophysical Union that he saw himself as an "enabler of science". After spending eight years building his one-man submersible, he had descended to the Challenger Deep with high-definition camera equipment – to capture images that Walsh and Piccard failed to – and brought back a sample of mud that has since been subject to scientific scrutiny.

Videos recorded by a remotely operated vehicle off the coast of California show the barreleye fish rotating its eyes within a transparent compartment at the front of its head, enabling it to see upwards as well as forwards.

– – →

Filmmaker James Cameron is one of only
three people to have made the journey
to the deepest spot in the ocean. On 25
March 2012, Cameron swaps notes with
Don Walsh – one of the other two – before
descending in his *Deepsea Challenger*
submersible.

Because the cost of building and operating a deep-sea vessel is colossal,
motivation must be strong. The search is always on for the secret to
the squid's disappearing act – the envy of the military – or the microbe
that produces an immensely valuable medicine. During shallower dives,
Cameron picked up traces of a chemical already being tested as a drug
for Alzheimer's disease, scyllo-inositol. This is found in the coconut palm,
but it turns out that some deep-sea crustaceans make it too. In fact, some
scientists think that discovering a potential drug in a crustacean or a
new species of marine bacterium may be better than producing random
chemicals in the laboratory, because the chemicals produced naturally
by these organisms have evolved for millions of years to be biologically
active, so they are more likely to interact with chemicals in our bodies,
and in the bugs that infect them, to fight disease.

Coral reefs may be some of the richest sources of new medicines in
nature, despite covering only a tiny fraction of the ocean floor. They are
thought to be home to around a quarter of the world's marine species
– a veritable treasure-trove stocked with enough naturally occurring
and novel compounds to keep the world's medicinal chemists busy for
decades. Recently discovered compounds produced by sea sponges that
live on corals may prove to be powerful new weapons in the war against
antibiotic resistance (*see* QUESTION 14: HOW WILL WE BEAT BACTERIA?).
Peter Moeller, a chemist at the Hollings Marine Laboratory in South
Carolina, and his team were studying disease-ridden coral reefs when they
found a sponge that seemed to be thriving in the midst of the death and
decay. Its secret, they realized, was an ability to keep invading bacteria
at bay by producing an antibacterial chemical called ageliferin. This
chemical could eventually be used to help bolster the effects of antibiotics
that are losing their fight against resistant bacteria.

Conditions in waters deeper than 3 kilometres (about 2 miles) are too cruel for corals, but scientists are also searching further down – including around black smokers – for chemicals with medical promise. We know that the pressure to survive in the extreme environment of the deep ocean drives organisms to strange evolutionary quirks; we only have to look to the barreleye fish with its transparent head for proof that the weird and wonderful exist down there. And such novelty is highly prized in the quest for new drugs. But there are other reasons to learn more about these extreme environments.

Scientists are interested in organisms known as "extremophiles", literally meaning "life forms that love extremes", such as the extremes of temperature, pressure, light and minerals peculiar to hydrothermal vents. Why? Because the fact that life exists in the most hostile environments on Earth implies that it may flourish elsewhere – somewhere out in the wider universe (*see* QUESTION 3: ARE WE ALONE IN THE UNIVERSE?). *Alvin* and other deep-sea submersibles have been kept busy in recent years by NASA-funded scientists studying microbes at hydrothermal vents in order to understand the limits at which life can survive. For, strange as it may seem, the darkest caverns of the deep sea may offer a perfect parallel for deep space.

Compounds collected from sea sponges hold the potential to fight super-bacteria that have developed resistance to antibiotics. But making copies of these compounds in the lab may be cheaper than extracting them straight from the source.

BIG DISCOVERY:
THERMAL SPRINGS SUPPORTING LIFE ON THE OCEAN FLOOR

"In February and March of 1977, we made a series of 24 dives in the deep submersible Alvin *on the 2.5-kilometer [1½-mile] deep axis of the Galápagos Rift…These dives enabled us to make direct visual observations of the area, to make small-scale physical measurements, and to obtain samples of fluids and related deposits at thermal-spring vents on the sea floor…In the course of our explorations, we discovered extraordinary communities of organisms living in the thermal vent areas at the rift axis…"*

From "Submarine Thermal Springs on the Galápagos Rift", published in the journal *Science*, 16 March 1979.

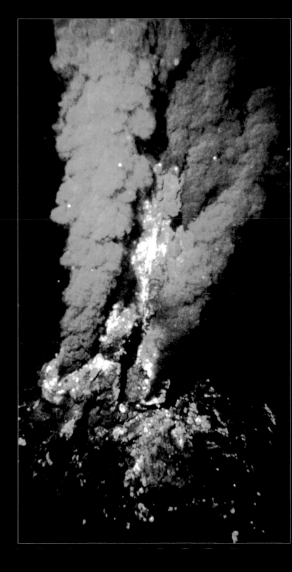

Hydrothermal vent systems snake - - ➔
across fault lines in the ocean
floor. Chimney-like structures
billow out plumes of "supercritical"
liquid – so hot, it is not quite water,
not quite gas – containing high
concentrations of minerals that
provide sustenance for life on the
ocean floor.

WHAT'S AT THE BOTTOM OF A BLACK HOLE?

The Sun is dying. Every second it uses up more of its hydrogen fuel, converting it into helium, light and other forms of electromagnetic radiation, and charged particles. With nothing to refuel, the Sun is slowly devouring the very thing that is keeping it going and is inching ever closer to the far-distant day of its – and our – demise. In about a billion years from now, having burned through a large fraction of its hydrogen, the Sun will begin to consume the rest at a faster pace. This elevated appetite for hydrogen will boost the Sun's brightness by around 10 per cent, with lethal consequences for our planet – the temperature on Earth will rise dramatically, incinerating its surface and boiling its water away. Life on land and sea will be wiped out. Four billion years later, our lifeless, sterile planet will take a further battering as the Sun bloats into a red giant star and surges out into the Solar System. The two planets nearest the Sun – Mercury and Venus – will eventually succumb to its fiery embrace, and as it balloons to 250 times its original size, the Earth may also be swallowed up. By then, the Sun's core will have contracted, raising the temperature enough to start converting helium into carbon and oxygen. But this helium-burning phase is nowhere near as stable as burning hydrogen – the Sun will pulse violently and these death tremors will eject its entire atmosphere out into space, plunging the Solar System into darkness. A tiny and ever-fading white dwarf star will be all that remains.

The ultimate fate of our Sun may sound catastrophic, but on a celestial scale it barely registers. The death of a more massive star can create one of the most exotic objects in the universe: a black hole. In this case, the star still becomes a red giant but then goes beyond the helium-burning phase and continues to fuse carbon, neon, oxygen and silicon in concentric shells so that its cross-section resembles the layers of an onion. The end comes when a dense iron core forms at the centre. Unable to support itself against its own weight, the core collapses and the star explodes in a supernova – one of the universe's most powerful and energetic phenomena – that briefly outshines billions of other stars.

What remains after the explosion of a sufficiently massive supernova is a black hole, which is currently best described by one of the most successful theories in the history of science: Albert Einstein's theory of general relativity. In 1915, the 36-year-old Einstein claimed that Isaac Newton, the godfather of physics, was wrong. In his original work on gravity, Newton had suggested that massive objects were capable of bending light. Einstein agreed, but said that light is bent twice as much. The Sun itself would help settle the debate. By far the most massive object in the Solar System, the Sun bends light enough for the deflection to be measurable. This means that when the Sun moves across our sky, stars close to it should appear in a slightly different position than normal due to the Sun's gravity bending their light on the way to Earth. Usually, however, the Sun is too bright to see this effect. Only during a total solar eclipse, when the Moon blocks out most of the Sun's glare, can such stars be seen and Einstein's prediction tested. In 1919, there was such an eclipse, and a photograph from it was used to show just how far starlight was deflected by the Sun. Einstein was vindicated and Newton's two-century-long reign came to an end (*see* BIG DISCOVERY: ECLIPSE CONFIRMS GENERAL RELATIVITY).

The key difference between the two great scientists was how they thought about gravity. Newton saw it as a force that pulled on the light and altered its path. Einstein's newly confirmed ideas were radical – he suggested that it wasn't the light that was bent but space itself. He combined the three dimensions of space and the one of time and said that the fabric of the universe was actually made up of four-dimensional spacetime. According to Einstein, this normally flat fabric is warped by the presence of massive objects like the Sun – just as a bowling ball would cause a dip in the centre of a tarpaulin sheet held tight at each corner. So when the light from stars is deflected by the Sun, it is simply because the starlight is following the local curvature of spacetime caused by the presence of the Sun. So, according to Einstein, a planet doesn't orbit a star because the star is pulling on it; rather, the mass of the star causes a dip in spacetime and the planet revolves around the lip of that dip.

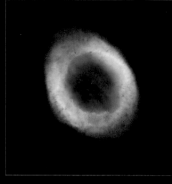

The Ring Nebula (M57) in the constellation of Lyra. Formed by the death of a star like the Sun, the star's material has been ejected out into space. A tiny white dwarf star, about the size of the Earth, remains at the centre.

← – –

The Crab Nebula (M1) in the constellation of Taurus. It is the remnant of a supernova explosion that occurred after the death of a massive star. Even bigger supernovas can leave black holes lurking at their centres.

When the iron core of a massive star collapses, the dip in nearby spacetime gets ever deeper as the core becomes increasingly dense. Eventually, the sides of the dip become so steep – the curvature of spacetime becomes so great – that the speed required to escape from the dip exceeds that of light. Einstein's previous work on special relativity had shown that it is impossible to travel faster than the speed of light, and so at this point not even photons (light particles) are able to escape the gravitational grasp of the collapsing core. Anything within this gravitational limit (known as the event horizon) is trapped forever, even light – which is why black holes are so named. So what happens to this perennially trapped material?

INFINITELY SMALL AND INFINITELY DENSE

According to general relativity, the iron core will continue collapsing until all of its original mass is concentrated in a single, infinitely small point – a singularity. Any material that subsequently falls into the black hole will be squashed down and added to the singularity. Because the singularity has an infinitely small volume, it is infinitely dense. And if you're wondering what an infinitely small or infinitely dense point might be like, you're not alone. To see the effect that infinities have on scientific equations, pick any number you like and try to divide it

The Sun bends the fabric of space around it, causing light to take a slightly different path when travelling to Earth from a distant star. Observations of a solar eclipse in 1919 showed this deflection to be precisely the amount predicted by Albert Einstein.

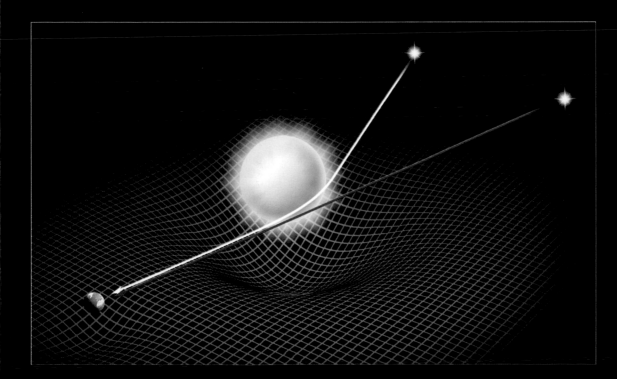

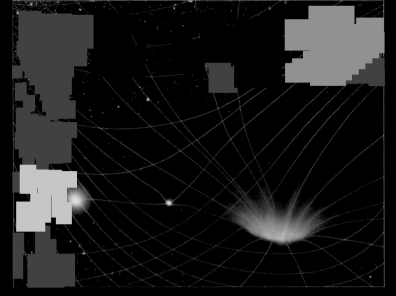

- - →

German mathematician Theodor Kaluza (1885–1945). Kaluza united Einstein's Theory of General Relativity with electromagnetism by invoking an extra, fifth dimension. Today, string theorists use a total of 11 dimensions to try and explain what happens at the bottom of a black hole.

← - -

The more massive an object, the more it warps spacetime. A neutron star (middle) – a dense remnant left over by small supernovas – creates a slightly bigger dip than our Sun (left). By contrast, a black hole (right) bends the space around it so much that not even light can escape.

by zero on a calculator – you get an error message. It is for this reason that researchers increasingly believe that a singularity is an incomplete picture of what really happens at the bottom of a black hole. It might simply disappear from the equations that describe a black hole if a more complete theory could be found that eliminates the infinities. Many researchers think that combining Einstein's general relativity with the other cornerstone of twentieth-century physics – quantum mechanics – might produce just such a theory.

Unfortunately, combining the two theories is easier said than done, as they are polar opposites. Quantum mechanics deals with the world of atoms, whereas general relativity explains the shape of space on the largest scales. Yet it is precisely their opposing natures that, when combined, makes them a good candidate for explaining the bottom of an object formed when large-scale space and matter is squeezed down beyond atomic size. If the theories could be combined, and the infinities disappeared, it might provide the true picture of what happens at the bottom of a black hole. It would also provide a powerful, one-size-fits-all theory capable of describing all phenomena across the universe, from the tiniest subatomic particle to the biggest galactic supercluster – a theory of everything. It is easy, then, to see why such a theory is so attractive, and a combination of the two might not just be wishful thinking either – unifications have happened in physics before. By drawing inspiration from these past successes, modern physicists are finding ways to eliminate the infinities that currently prevent a proper understanding of black holes.

LIGHT AND GRAVITY UNITE

One such unification happened in the nineteenth century when Scottish physicist James Clerk Maxwell realized that electricity and magnetism are intertwined. In a landmark publication of 1865, Maxwell unified both phenomena by establishing that they were "affections of the same substance" – in modern terms, aspects of the same thing, now known as the electromagnetic field. Around the same time, Maxwell also published four equations describing the interaction between electric and magnetic fields, currents and charges. These have since become known as Maxwell's equations. Then, in 1919, the year of the solar eclipse that provided evidence supporting Einstein's theory of general relativity, the German mathematician Theodor Kaluza began to wonder whether he could achieve further unification. Einstein had succeeded in explaining gravity by describing it as a warping of the three dimensions of space combined with the one dimension of time. It occurred to Kaluza that he might be able to combine Maxwell's equations of electromagnetism with Einstein's equations of general relativity by invoking an extra dimension of space.

INSIDE EXPERT: WHY ELEVEN DIMENSIONS?

"Subatomic particles have an intrinsic property called 'spin' which separates them into two classes. Particles that have spin in whole-number multiples (0, 1, 2…) are called bosons, and those that have half-number multiples of spin (½, ³⁄₂…) are called fermions. Particles that make up atoms, like electrons and quarks (both with spin ½), are therefore fermions, while force-carrying particles, such as the photon (with spin 1) and the graviton (spin 2), are bosons. It is believed that the spin of truly elementary massless particles cannot exceed 2.

The different nature of fermions and bosons means that, conventionally, we have had to use different equations to describe them. However, if we use a theory that is 'supersymmetric', we can swap the fermions and bosons while keeping the equations the same. This supersymmetry, first proposed in the 1970s, places an upper limit on how many dimensions can exist – 11 (10 space, 1 time). Otherwise, after curling up the extra dimensions, as in the Kaluza–Klein theory, there would be massless particles with spin greater than 2.

M-theory describes the movement of two-dimensional objects (membranes) moving in such an 11-dimensional world. If one of the dimensions is a circle, these membranes can wrap around it. If the radius of the circle is sufficiently small, this gives the appearance of a one-dimensional object (a string) moving in 10 spacetime dimensions. This is what superstring theory originally described.

M-theory is able to incorporate quantum mechanics and general relativity into the same framework and may one day allow them to be unified into an all-embracing theory that could, in particular, explain what happens at the centre of a black hole."

Michael Duff FRS,
Abdus Salam Professor of Theoretical Physics,
Imperial College London.

When he did the mathematics, he found that, in a five-dimensional universe (four of space and one of time), relativity and electromagnetism could be unified into one set of equations, and he duly published his findings in 1921.

As beautifully elegant as Kaluza's solution was, there was a glaring problem: if the universe really had an extra, fourth dimension of space, why do we see only three? It took another five years for a possible answer to emerge, and it came from Swedish physicist Oskar Klein. The 32-year-old suggested that perhaps the extra dimension was real but that it was curled up so small that we don't observe it. Imagine an ant crawling around the branch of a tree. From far away the branch appears flat and two dimensional. Only from an ant's perspective, when you get right up close, would you be able to see, and have access to, the extra dimension of its circumference. Klein suggested that if Kaluza's extra dimension was curled up small enough, we wouldn't notice it. To remain undetected, he claimed it would need to be curled up on the scale of 1.6×10^{-35} metres (6.4×10^{-34} inches) – 0.0000000000000000000000000000000000016 metres, or 0.00000000000000000000000000000000000064 inches – the smallest meaningful distance in quantum physics, known as the Planck length.

For their combined efforts on the problem, this work is now referred to as the Kaluza–Klein theory. In principle at least, Kaluza and Klein had managed to unite general relativity with the seemingly unconnected theory of electromagnetism without any infinities to complicate matters. This is exactly what physicists trying to combine general relativity with quantum mechanics were striving for. Yet, in the decades that followed,

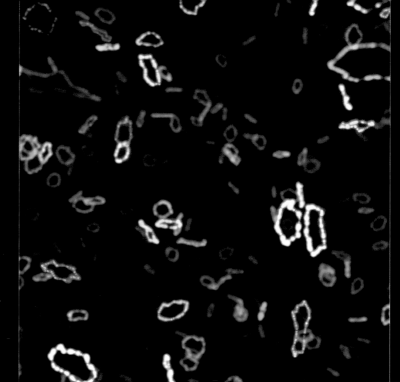

The tiny strings that string theory suggests vibrate in 10 dimensions to produce the fundamental building blocks of nature. When the theory tries to combine general relativity with quantum mechanics, the infinities that have plagued physics for a century disappear.

the Kaluza–Klein theory was largely forgotten because the extra dimension it invoked couldn't be reconciled with reality. Physicists continued searching for ways to unite quantum mechanics and general relativity, but they continued to face the same problem: infinities kept creeping into the equations. But in the 1980s, Kaluza and Klein's idea enjoyed a revival and the search for unification took a promising turn. A new idea – superstring theory – used a variation of the extra dimensions idea to offer a way to combine the two theories, providing a possible explanation for what happens at the bottom of a black hole.

VIBRATING STRINGS

The "string" part of superstring theory comes from the fact that it postulates that the smallest, most fundamental building blocks of nature are strings that vibrate across multiple dimensions. In the same way that strings on a musical instrument can vibrate in different ways to produce different musical notes, the different vibrations of these superstrings produce the different subatomic particles that make up everything around us. The "super" part refers to the fact that the theory incorporates another theory called supersymmetry, which limits the number of dimensions in which the strings can vibrate to 10 – nine of space and one of time. It is suggested that we don't experience the extra six dimensions of space because, as in the Kaluza–Klein theory, they are curled up so small that we can't see them. Curiously, however, the total maximum number of dimensions permitted by supersymmetry is not 10 but 11. So why are the strings permitted to vibrate in only 10 dimensions? This mystery was resolved when it was realized that the one-dimensional strings moving in 10 dimensions of space and time are really two-dimensional membranes wrapped around the 11th dimension. Superstring theory has thus been extended into 11-dimensional M-theory (*see* INSIDE EXPERT: WHY ELEVEN DIMENSIONS?). When physicists tried to use this 11-dimensional world of M-theory to apply quantum mechanics to general relativity, they found something remarkable: the infinities in the equations disappeared.

There is, however, a catch. There are an enormous number of ways in which you can curl up the extra seven dimensions, and each configuration would produce different sets of subatomic particles. It isn't yet known which configuration of curled-up dimensions would give rise to the set of subatomic particles found in our universe. Until a version of M-theory is devised that can predict the existence of things we already know about, there will always be questions about whether the extra dimensions really exist. By invoking extra, unseen dimensions, M-theory may make quantum mechanics compatible with general relativity in theory, but we can't yet know if it is an accurate description of the universe we live in. Critics of M-theory also ask why, if it is a true reflection of things, nature has selected our particular M-theory configuration from the numerous possible alternatives? One explanation is that ours isn't the only universe. If an infinite number of universes exist, then all the different configurations would exist somewhere. If so, it isn't a surprise that we live in a universe where the configuration gives rise to just the right combination of subatomic particles and forces that can form galaxies, stars, planets and people (*see* QUESTION 8: ARE THERE OTHER UNIVERSES?).

Despite these concerns, M-theory remains one of the most fashionable topics in modern physics, with thousands of researchers working on it around the world. The attraction is clear – it seems to offer a way to eliminate the infinities that have plagued physicists for a century. If it can truly unite the seemingly disparate theories of quantum mechanics and general relativity, then singularities could well be consigned to the history books. But much more than that, it could provide the holy grail of physics: knowledge of the fundamental inner workings of the universe – a theory of everything. And we'd have an answer to the mystery of what really happens at the bottom of a black hole.

A total eclipse of the Sun. Taken by British astronomer Arthur Eddington on 29 May 1919 from the tiny island of Principe, the image shows the Moon passing in front of our star. Observations of this eclipse allowed Eddington to vindicate Einstein's ideas about gravity.

BIG DISCOVERY:
ECLIPSE CONFIRMS GENERAL RELATIVITY

It had rained for 19 consecutive days on the tiny island of Principe just off the west coast of Africa. British astronomer Arthur Eddington was becoming increasingly frustrated – the 1919 total eclipse of the Sun had started without him. Totality, when the Moon completely blocks the main disc of the Sun, was scheduled to last for 420 seconds, but 410 seconds had already elapsed with the clouds preventing any observation. Then, just as it looked like his opportunity had passed, the clouds cleared and Eddington managed to take the single photograph that supported the predictions made by Albert Einstein in his general theory of relativity.

The reasons behind Eddington's expedition were rooted in the recently ended First World War. German-born Einstein had published his theory in 1915, news of which had made its way to England despite the ongoing conflict. Eddington, a Quaker and pacifist, managed to avoid conscription to the front line and likely punishment as a conscientious objector thanks to a letter from the Astronomer Royal, Sir Frank Dyson, who argued that England couldn't afford to lose one of its most promising astronomers on the battlefield. Instead, Eddington was recruited to travel to Principe in 1919 to test Einstein's predictions for how much the Sun's gravity bends light. His photograph backed up the theory which, today, stands as one of the most-tested theories in science. Among other things, it helps to explain how a black hole is formed when a massive star explodes at the end of its life.

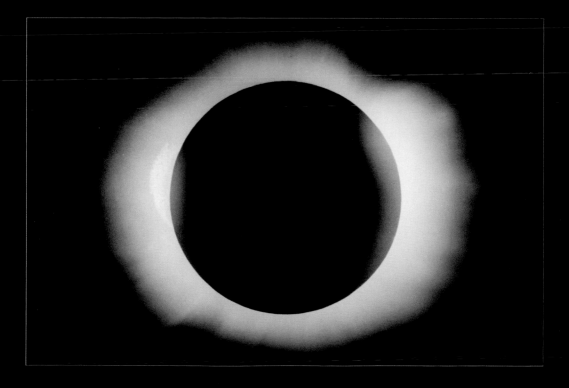

CAN WE LIVE
FOREVER?

On 8 January 2013, Brooke Greenberg turned 20 years old. For her, though, there were no wild parties or coming-of-age ceremonies because, however you look at it, Brooke is still a child. And not in the metaphorical sense – she has literally hardly aged since the day she was born. Delivered at Sinai Hospital in the US city of Baltimore, Brooke suffered a number of unexplained medical emergencies in her early childhood from which she miraculously recovered. There were seven perforated stomach ulcers and a stroke that left no apparent brain damage. At the age of four, she developed a brain tumour that put her to sleep for 14 days. Her parents thought she would die; yet she awoke of her own accord, with the tumour mysteriously vanished. Brooke looks more 20 months than 20 years old: her brain has changed little since infancy, and many of her teeth are still baby teeth; only her hair and nails seem to grow normally. And we still don't know why, although scientists think it may be "disorganized development", with different parts of her body developing out of sync.

On the opposite side of the spectrum are those with diseases like progeria or Werner's syndrome that mimic the effects of accelerated ageing: growth failure, loss of body fat and hair, wrinkled skin, stiff joints, hip dislocations, and increased incidence of stroke, heart disease, diabetes and cancer. All of these conditions are very rare: progeria affects 1 in 8 million births, Werner's syndrome 1 in 10 million, and just a handful of people worldwide have symptoms similar to Brooke's. We don't have a cure for any of these, but what we're learning from all of them is that ageing is not fixed but is actually quite flexible.

Evidence from the animal world confirms that this is not such a misguided notion. The mayfly famously lives for only a day, but the great Galapagos tortoise Lonesome George lived for about 100 years and turtles are even known to live to 200. Yet the longest a human has survived is 122 years and 164 days. So what's the reason for this?

DISTRESSED GENES

In the 1930s, US scientists Barbara McClintock and Hermann J. Muller, two of the greatest ever geneticists, found that the chromosomes (the coiled-up form of DNA, the cell's genetic material) have a special region at their ends that keep them stable. Often described as being like the plastic caps on the end of shoelaces, these regions – called telomeres – ensure that the whole DNA strand is copied when a cell divides. Fast-forward 50 years and Australian-American Elizabeth Blackburn and Canadian-American Jack W. Szostak discovered something astounding: shortening the telomeres causes premature ageing in yeast and human cells – a discovery that earned them a share of the 2009 Nobel Prize in physiology or medicine.

Lonesome George, the giant Pinta tortoise of the Galápagos Islands, Ecuador. He first encountered humans in 1971 and died in June 2012 at an estimated age of 100.

← – –

Subsequent experiments on mice and other animals have shown that they live longer if they are genetically modified to have longer telomeres. So many scientists believe ageing is at least partly due to these preprogrammed timers, ticking down inside each of our cells. Each time a cell divides, it loses the ends of its telomeres – the more divisions, the shorter the telomeres, until eventually a cell can divide no more. However, some cells can keep dividing indefinitely – stem cells or cancer cells, for instance. The difference is due to an enzyme called telomerase, which repairs the telomeres, keeping them from getting shorter when a cell divides. Many human cells lack telomerase, which explains why they can divide only a certain number of times – cells from a newborn baby are able to divide 80 to 90 times, whereas those from a 70-year-old usually manage just 20 to 30 divisions, functioning less efficiently until they eventually die.

Telomere length is associated with longevity and resistance to disease in long-lived humans, and telomere problems are involved in progeria and Werner's syndrome. Interestingly, Brooke Greenberg's telomeres seem to be shorter than those of normal infants and those of people her own age. Researchers think that this is probably the result of her prolonged infancy – the rate of telomere shortening is known to be a lot higher in infants than in adults. However, before we think that telomeres are the master switch for ageing, other findings suggest that ageing is more complex. For one thing, mice with no telomerase – and therefore no telomere repair mechanism – don't seem to suffer any particular effects on ageing for several generations.

Nevertheless, telomere shortening also ties in with what many scientists believe to be the cause of ageing: wear and tear. As you get older, your body is less able to maintain itself, the mechanisms that repair damage stop working and things just go wrong. Much of this is the result of DNA damage and genetic mutations, which might be due to a wide variety of factors, such as environmental pollution, radiation and oxygen free radicals (highly reactive oxygen molecules). It makes a lot of sense to think of ageing as a fundamental conflict between maturation and deterioration, although you might then wonder why we have evolved to age at all. For any organism the primary goal in life is to procreate, and in 1957 US scientist George C. Williams proposed that ageing might be a trade-off: a stronger, fitter body for procreating earlier in life at the cost of later deterioration. This fits with evidence that mutations in some of our genes increase fertility but also the risk of cancer (*see* QUESTION 15: WILL WE EVER CURE CANCER?); they might improve your chances of having children, but are bought with a huge price tag and a killer credit card bill to pay later. That cost implies some kind of budget. So, in 1977, the British biologist Tom Kirkwood suggested that the body has a limited amount of energy available and must decide how best to spend it: general maintenance and repair or better fertility. Any change to boost fertility has a detrimental effect on the body's ability to repair itself, and vice versa. In the long run, it makes more sense to procreate rather than build a body that could last forever – certainly so in our brutal past when there was always the chance that some predator, disease or accident could kill you at any time.

EAT LESS, LIVE LONG AND PROSPER

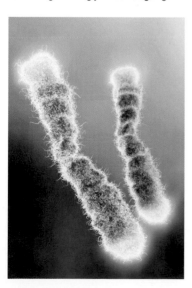

If chromosomes were shoelaces, the telomeres would be the "plastic tips" at the ends. Telomeres prevent the DNA at the ends of the chromosomes from degrading or sticking together. Their shortening is strongly linked to ageing.

In the 1930s, the USA suffered the Great Depression. Concerned about the effects of starvation, scientists started experimenting on rats. The results were unexpected: rats that ate less food didn't get sick but lived on, sometimes up to 50 per cent longer than those on a normal diet. This remarkable effect has been confirmed in yeast, fruit flies, roundworms and mice. In many organisms, restricting calorie intake leads to phenomenal gains in lifespan – up to 10 times in some nematode worms, the equivalent of 1,000 human years. But would this work in humans?

Unfortunately, there's still no clear-cut answer. Two studies on rhesus monkeys have come up with conflicting results. The first, published in 2009 by researchers at the University of Wisconsin, found that calorie restriction did indeed work: over 20 years, monkeys on a calorie-restricted diet lived longer and had lower incidence of age-related disease than monkeys on a normal diet. But a 2012 study conducted over 23 years by researchers at the US National Institute on Aging concluded the opposite: that calorie restriction didn't extend life. They found that, irrespective of the age at which the monkeys started their diet, those fed 30 per cent fewer calories didn't live any longer than the control animals on a normal diet. However, both studies did note a significant improvement in the health of the calorie-restricted monkeys, with delayed onset of age-related diseases such as cancer and cardiovascular disease and, in the Wisconsin calorie-restricted monkeys, a complete absence of diabetes. It would be unrealistic to do studies like these in humans, but it hasn't stopped many people from trying calorie restriction anyway, and scientists are monitoring those who voluntarily submit themselves to this in the hope of gathering meaningful data.

We also know, of course, that being thin does not necessarily equate to a longer life, so why should eating less slow ageing? It seems to fly in the face of our best theories: surely a lower food intake means you have less essential resources to maintain your body? Scientists initially thought that calorie restriction simply reduced the number of chemical reactions in the body and, therefore, any cell damage that occurs naturally as a result of those reactions. Yet it is so effective and relatively fast-acting that that can't be all there is to it. What scientists are discovering is a link back to that balancing act we discussed earlier: a fitter body or longer life.

SWEET SIXTEEN AND THE GRIM REAPER

It's the 1990s and Cynthia Kenyon at the University of California, San Francisco is looking for genes that affect ageing. It's unfashionable - most scientists believe altering ageing, let alone achieving immortality, to be impossible – so she has only one graduate student, Ramon Tabtiang, to help her. They try anyway, systematically breeding mutants of the nematode worm *Caenorhabditis elegans*, searching for any that live longer and stay healthy. One day, Ramon walks into Kenyon's office and says, "Guess what? They're not dying".

That mutant worm lived twice as long as any normal worm – all from turning off just one gene (called DAF-2) out of the 20,000 that make up the worm's entire genome. Kenyon dubbed the DAF-2 gene a "grim reaper gene" because its normal function seems to stop an animal from staying young. Its discovery unveiled the first molecular pathway known to regulate ageing in the body. The grim reaper gene controls a hormone involved in growth, which in turn affects a cascade of genes and molecules along the complex molecular pathway. In particular, activity of the grim reaper gene suppresses the activity of another gene called DAF-16, which Kenyon nicknamed "sweet sixteen" because boosting DAF-16 activity leads to longer life.

Although first discovered in a nematode worm, this pathway is essential in all animals, including humans. Studies of Japanese, Ashkenazi Jew and other populations who are exceptionally long-lived, have revealed a number of mutations that impair the grim reaper gene, while several human equivalents of sweet sixteen have been found in Italian, German, Chinese, Californian and New England people who've lived to 100 years and beyond. Scientists think that genetic mutations like this make the body better or worse at protecting itself from damage; the actual effect depends on the environment, particularly the amount of food available. When food is plentiful and stress levels are low, growth and reproduction are favoured

- - →

The Great Depression of the 1930s saw many people out of work and unable to afford essentials such as food, leaving US scientists worried about the impact of starvation.

INSIDE EXPERT: IS AGEING A DISEASE?

"The predominant view of medicine is that ageing is not a disease but a natural process, part of the life cycle. But if you look at ageing from the point of view of a biologist rather than a clinician, it's actually a process of catastrophic deterioration that leads to pathologies that you die from. Most of the diseases that people in the developed world die of today are actually components of this broader process. Numerically, most fatal cancers that appear are in the elderly and are symptomatic of ageing.

The predominant theory has long been that ageing is essentially a wear and tear process. In particular, there accumulates damage in the proteins, nucleic acids, etc. essential to life. Your body can defend itself against this via maintenance processes (e.g. repair and turnover) and how well it does so determines the rate of ageing. A lot of the theories of ageing, such as the free radical theory or the shortening of telomeres, fall under the umbrella of this damage/maintenance paradigm. Although it seems very common sense, there are things that it can't explain. For example, many recent studies have shown that altering maintenance levels doesn't have the expected effect on ageing. Also, reducing nutrient availability (dietary restriction), which should reduce energy for maintenance processes and speed up ageing, actually has the opposite effect and extends lifespan.

One new alternative is the hyperfunction theory, proposed recently by Mikhail Blagosklonny [Professor of Oncology at the Roswell Park Cancer Institute, New York]. It suggests that ageing results from the running-on of the body's developmental programmes into later life. This is based on recent scientific discoveries, which have shown that the pathways that control cell growth and proliferation in particular also seem to control ageing. The hyperfunction theory proposes that evolution acts on hormonal and cellular signalling pathways, optimizing them to maximize reproductive success. But after reproduction, pathway activity is no longer optimal and often too high. This leads to diseases of hypertrophy (increase in organ or tissue volume), hyperplasia (cell proliferation) – including cancer – and atrophy (wasting away). And that's what tends to kill you in late life. It's a good theory because it provides an integrated picture of what ageing is, combining both the evolutionary and mechanistic biology."

Dr David Gems,
UCL Institute of Healthy Ageing, London, UK

over maintenance, but under harsher conditions the body instead prioritizes cell protection and maintenance, guarding against environmental stresses and extending lifespan. Live to procreate another day, so to speak, when conditions might also favour the survival of your offspring.

A PILL FOR ALL ILLS

These findings are a hint that each animal has within it the ability to live much longer than normal. Perhaps by manipulating the genes controlling pathways that affect how the body responds to environmental stresses, we can trick the body into thinking it is under threat, so that it goes into a "protect and maintain" state. As well as the DAF-2/DAF-16 pathway, another such pathway is TOR, manipulation of which helps yeast, flies, worms and mice to live longer. The interesting thing about this is that we have a drug, sirolimus (also known as rapamycin) that works on it. We know this drug can extend life in mice, and it's already approved for use in the US and UK – but as an immunosuppressant, the type of drug given to transplant patients to weaken their immune system so that the donor organ isn't rejected. So it's not ideal as an anti-ageing treatment. Attempts to find drugs targeting other, similar pathways have also been problematic, none more so than those involving the sirtuin genes. Despite thousands of scientific papers published on this group, there's still much debate about whether they extend lifespan or not. In 2004, there were claims that a molecule found in red wine boosted sirtuin activity, extending life in yeast. The link with red wine has now been disproved, but that's not to say that sirtuins should be written off altogether. Certainly there is evidence that mutations in sirtuin genes make an animal healthier during its life – lowering cholesterol and reducing loss of muscle mass, cognitive decline and incidence of type 2 diabetes – although this hasn't helped sirtuin mutant mice, for instance, live much longer than normal.

As a result of all these discoveries, we're starting to think of ageing not as a fact of life but more like a disease that can be treated like any other. This is an important shift in mindset for two reasons. First, many diseases are primarily diseases of ageing – type 2 diabetes, cancer, osteoporosis and cardiovascular disease, for example – so successfully tackling ageing would allow us to combat many diseases at once. The second reason is treatment and regulation. Medical and drug licensing authorities currently recognize only treatments that target a disease, so unless ageing is a disease, any anti-ageing drug wouldn't be medically approved. Then there's the vast number of purported anti-ageing products already on the market, at best regulated as cosmetics or dietary supplements, at worst not regulated (and with no requirement for testing) at all. Reclassifying ageing as a disease would bring all of these products under the umbrella of proper evidence-based medicine.

Brooke Greenberg at the age of 16. Brooke has barely aged physically since she was born over 20 years ago. Her unusual condition continues to mystify doctors and scientists.

LIVING LONG AND WELL

Some scientists believe life extension is a question of "when" not "if". Thanks to advances in health and medicine, we are living longer than ever before: between 1970 and 2010, average life expectancy worldwide rose by 11 years for men and 12 years for women – about 2.5 years each decade. The idea of immortality, or at least living for a very, very long time, is now less fantasy, more actual possibility. If you can stay alive for 20 years, say, until the next medical advances, maybe you could keep on going for another 20 years, and another. But that's not to say we can

live forever; there's always the possibility of succumbing to disease or an accident. Indeed, the laboratory animals that have been engineered to live longer still eventually die of age-related diseases. To quote British gerontologist David Gems of the UCL Institute of Healthy Ageing, London, "decelerated ageing is not expected to change the overall, lifetime risk of disease; rather, it reduces disease incidence at any given age".

Even if we could extend life continually, living forever does not equate to living well forever. As Gems has put it, "People aren't afraid of dying, they're afraid of having to have someone else take them to the toilet". The World Health Organization says that, between 2000 and 2050, the number of people over 60 years old worldwide will rise to 2 billion, with 400 million over 80 (*see* QUESTION 19: HOW DO WE SOLVE THE POPULATION PROBLEM?). Much of ageing research therefore concentrates on healthy ageing, accepting that we will continue to prolong life and focusing on what we need to do to ensure that those later years are as comfortable as possible. The 20-year-old Brooke Greenberg remains young but in full-time care. She still wears nappies and has to be fed by a tube because her oesophagus isn't developed enough. But she likes giggling and still enjoys trips in a stroller with her mother. What matters is not the quantity of life, but the quality.

19

HOW DO WE SOLVE THE POPULATION PROBLEM?

A a species, *Homo sapiens* is extraordinarily successful. Not only has it survived for hundreds of thousands of years, it has also become the dominant land-dwelling mammal. The International Union for Conservation of Nature and Natural Resources (IUCN) which classifies species according to conservation status, includes *Homo sapiens* in its category of "least concern". In other words, there is almost no chance that humans will become extinct any time soon. We have colonized every corner of the Earth, and even managed to escape from it. As the IUCN notes, "a small group of humans has been introduced to space, where they inhabit the International Space Station". Yet it is the success of our species that is at the root of many of its problems. As the human population continues to grow exponentially, as city life becomes more cramped, as food, water and fuel supplies dwindle, we will need all our ingenuity, in medicine, agriculture and technology, to try to make our stay on this planet a little more comfortable for a little longer.

The population problem can be looked at in one of two ways: either there are too many people or there are too few resources. If, as Stanford University professor Paul Ehrlich believes, there are too many people, then there must be an optimum number. Ehrlich, a butterfly expert, became a controversial figure in 1968 when he published a book called *The Population Bomb*. In it, he claimed that the battle to feed humanity was already over and that death rates would start to increase within a decade. He was wrong but continues to claim that there is an optimum number of people – around two billion – that the planet can sustain. At seven billion and rising, this "optimum" has already been far surpassed. On the other hand, we might consider that our soaring population could be sustained simply by making sure that the resources we have are utilized and distributed more carefully. At present, we are consuming renewable resources such as crops, timber and fish faster than the Earth is capable of replenishing them. It takes about 18 months to regenerate the resources that we use in one year. We are doing the ecological equivalent of paying by credit card and hoping the bill never arrives.

Whether or not there are too many people, it is this ecological debt that we need to address by making industries like farming, forestry and fishing more sustainable and by saving resources. Even resources we might never have thought of as such are disappearing. Soil is one of the most valuable resources we have – we need it to hold water, grow food and absorb carbon from organic matter to help regulate our climate (*see* QUESTION 0: WHERE DO WE PUT ALL THE CARBON?). In 2012, John Crawford the University of Sydney in Australia predicted that topsoil would run out in around 60 years. This sounds far-fetched until we appreciate that topsoil is not simply the upper portion of the soil, it is the microbe-rich layer packed with nutrients that feed our crops. When soil becomes badly degraded through unsustainable farming, the topsoil is stripped Crops fail and water drains straight through, pouring back out into rivers and oceans and contributing to sea level rise, which is already devouring our coastlines and limiting the land available to house our ever-expanding population.

We share the planet and its resources with other species, some more numerous than ourselves. Our coexistence brings us into conflict; humans have already killed off enough species – including as many as half of all amphibian species – to constitute what some scientists are referring to as the Earth's sixth mass extinction. (The fifth mass extinction – best known for the extinction of the dinosaurs – occurred about 65 million years ago.) But it works both ways. Despite advances in medicine in the last few centuries, we are in an ongoing arms race with bacteria (*see* QUESTION 14: HOW WILL WE BEAT BACTERIA?) and viruses. It may be only a matter of time before the next big pandemic strikes, spreading all the more quickly due to the dense concentrations of humans in cities. It is worth remembering that the 1918 flu pandemic at the end of the First World War killed tens of millions of people – more than the war itself, and possibly more than died from the Black Death in the Middle Ages. It turns out that the 1918 flu virus is similar to modern viruses found in some bird flu cases, such as in the 1997 H5N1 outbreak that started in

A busy day in 2010 at the Nagashima Spaland amusement park in Kuwana, Japan. With a population density that is among the top 50 in the world, there are 337 people per square kilometre (873 per square mile) living in Japan. In the highly developed Macau region of China, there are 19,610 per square kilometre (50,790 per square mile).

Hong Kong. But while the 1918 flu could spread from person to person, H5N1 could not. While this hints that we may have had a lucky escape in 1997, it also reminds us that we may not be so lucky in future.

Nevertheless, the IUCN's Red List of endangered species assures us that, apart from localized declines due to disease and natural disasters, *Homo sapiens* is currently facing "no major threats" and that "no conservation measures are required". So for now, the population trend is upward. Ehrlich, for his part, has advocated birth control, sterilization, abortion and, where necessary, government intervention to control the human population. China, which in 2012 was home to nearly a fifth of the world's population, has been experimenting with population control since the late 1970s. Amid continuing controversy, estimates suggest that the one-child policy – brought into force in 1979 though relaxed in recent years – may have prevented hundreds of millions of births, although the country's population continues to increase. Meanwhile, the populations of other developing nations are rising faster than expected. In the 2000s, the United Nations estimated that world population would peak and level off at around nine billion, but in 2011 it revised its predictions based on higher than expected birth rates in Africa and lower than expected death

An emergency flu ward set up by the US army in Kansas during the 1918 influenza pandemic. The 1918 strain, also known as Spanish flu, infected a fifth of the world's population.

rates from HIV/AIDS. And while the populations of developed nations stagnate and grow older as people choose to have fewer children, it seems we cannot curb our in-built desire to start a family – one study suggests the symptoms of "baby-longing" are overpowering (*see* BIG DISCOVERY: BABY-LONGING). More practically, many African people see a large family as a means for support in their old age – and a brighter future. However, modernization may change attitudes and behaviour. Studies carried out in China and Africa in recent decades show that women who watch television have fewer children, even when other factors such as education and wealth are accounted for, and a 2011 study on this seemingly strange association suggested that soap operas, in particular, were changing traditional attitudes to marriage and family.

DOWN AND OUT

Predictions suggest that there will be around ten billion people by 2100, with more than a third of them living in Africa. Where are we going to put everyone? There is a spectrum of possible living solutions, which range from innovative to science fiction. At the innovative end of the scale are more compact homes that make use of every bit of available space. Some housing plots in Tokyo, Japan, for example, are no bigger than a car parking space. Among architects, there is a trend for sustainable pod homes – prefabricated houses made from recycled materials and capable of generating their own energy through solar panels. These "microhomes" also save on heating energy simply by virtue of being small. In reality, though, with more than half of us living in cities, individual microhomes may not be practical for urban living. Tower blocks already maximize the living space available on limited plots but have failed to live up to the dream of "cities

A slum settlement in Nairobi, Kenya in East Africa. In 2008, according to The World Bank, 1.29 billion people were living in extreme poverty (below $1.25 a day). Three-quarters of these people were living in South Asia and Sub-Saharan Africa.

← - -

Life on Mars? In 1973, Carl Sagan proposed making Mars more habitable by spreading soot on its polar caps – thereby reducing its reflectivity and triggering a change in climate to produce a milder, waterier world. But until humans permanently alter the red planet's atmosphere, breathing will remain an issue, meaning living and working indoors in modified atmosphere buildings.

in the sky". Being confined to a small box of a home has its disadvantages, not least the difficulty of escaping in an emergency when that box is many stories above the ground. Another option, occupying the middle ground between innovation and science fiction, is subterranean living. In early 2013, it emerged that a new Singaporean agency operating under the country's Building Construction Authority was carrying out underground surveys with the aim of eventually creating underground living spaces in giant caves. Plans to extend Singapore's Nanyang Technological University four levels down already exist.

At the science fiction end of the scale, there exists the tantalizing possibility of living on Mars. In 2013, a Dutch company, Mars One, announced its plans to establish a colony on Mars by 2023, a lofty goal considering not one person has yet stood on the surface of the planet. It plans to fund the project through a global reality show and corporate sponsorship. NASA, the US space agency, is slightly less ambitious; in 2012, its chief, Charles Bolden, suggested that the first manned spacecraft could reach Mars by around 2035. The challenges of making Mars suitable for human habitation are formidable: a thin, unbreathable atmosphere, freezing temperatures, high levels of radiation and weak gravity (38 per cent of the Earth's). Prefabricated habitats could be sent to Mars to provide a habitable environment on the planet, but the psychological effects of spending long periods in space just to get to Mars and then adapting to an alien world will be just as challenging.

FEED THE WORLD

Right now, though, there is a more immediate problem: how to ensure there is enough food, water and energy to go around. Making food and water readily available to everyone is already problematic, fossil fuels are running out (*see* QUESTION 10: HOW DO WE GET MORE ENERGY FROM THE SUN?) and the future will require big changes in the way we use our resources to meet growing demand. For a start, we will have to make some dramatic changes to our diets to reduce the pressure on farmland. The consequences of carrying on as usual are alarming: the amount of water consumed by livestock farming (which uses more than growing crops) would, at levels required to feed an exploding population, leave us without enough to drink. Scientists have calculated that, by 2050, we will need to reduce the amount of calories we get from animal products from 20 per cent to 5 per cent and get more of our protein from plants to ensure there is sufficient water. Already, groundwater has to be pumped up from underground to supply the vast volumes that intensive farming demands. Estimates from 2008 suggest that 147 cubic kilometres (35 cubic miles) of groundwater are removed each year – enough to fill a swimming pool the size of Paris. At the same time, the overcrowding of cities means that water is being drawn from further outside urban areas in order to cope with increasing demand. By the middle of the century, water shortage may be a very real problem for more than half of the world's population.

For committed carnivores, it may come as little consolation to know that scientists are working on growing meat in the laboratory. In 2012, Mark Post, a physiologist at Eindhoven University in the Netherlands, announced his laboratory would serve up the first artificially created burger by the end of the year. Distinct from meat substitutes such as mycoprotein, which is grown in tanks of fermenting mould, a laboratory burger would contain real animal cells – Post's were muscle cells cultured in a petri dish. Although each cell would have a lineage stretching back to a real animal, the cells would not naturally form what we would recognize as a piece of meat and would have to be mixed with fat cells grown separately to make a juicy burger. The neat thing about this technology is that it makes use of more of the nutrients that are fed into the process – by contrast, cattle-grazing is wasteful because most of the nutrients in plants are used up by the animals instead of going straight into our food.

Post's promised burger did not materialize on schedule and there remains some way to go until the commercial aspects are smoothed out, but what is perhaps of greater concern to food manufacturers is the consumers' perception of these new technologies. Genetically modified foods (*see* INSIDE EXPERT: CAN GM CROPS HELP FEED THE WORLD?) were at first roundly rejected, despite the wealth of benefits that GM crops

– – →

Genetically modified canola (rapeseed) crop growing in Victoria, Australia. Canola is often sold as vegetable oil, with demand steadily increasing since the 1970s. The GM crop has genes that make it resistant to a common herbicide.

INSIDE EXPERT:
CAN GM CROPS HELP FEED THE WORLD?

"We have a global population that is growing, we are concerned about climate change and consumption is increasing as some developing countries get richer. We need to develop technologies that will allow us to produce food resources in a way that is adaptable to changing conditions and, because we have a lot of people to feed, we have to try to get the most with the least resources. If we have efficient use of meat and efficient production of grains, we will be able to preserve wildlife. If we don't, we'll destroy all the forests and climate change will be worse. GM food is a technology that allows much faster adaptation to climate change and much faster development of new genetic varieties. In the US, Brazil and Argentina, GM food has already made a huge difference – the production of soybean in Argentina has tripled because of GM food. Thus far GM has been adopted partially, but not fully. If the rest of the world adopted GM corn and soybean, it could increase supply by 20%. If it was adopted with rice and wheat, we would have lower prices and less land in farming. Of course, GM poses risks, what technology does not? But so far it has had a great track record. In Europe, there are a lot of political and economic obstacles to the use of GM foods. But not using it is like not using electricity because we are afraid of electrocution. I think it's a very simple issue: we have a technology that's much safer than pesticides and can complement organic foods because a lot of chemicals can be replaced. So let's give it a try. If we are afraid of all the little things, then we will be destroyed by the bigger issue. The challenges are investing in research, developing mechanisms to apply this research and transferring it to poor farmers, but I think humanity can address these challenges."

David Zilberman,
professor of agriculture and resource economics,
University of California, Berkeley, US

offer, such as resistance to disease and drought, and bigger yields. The understandable feeling of distaste for food that is perceived of as fake or unnatural persists. But as farmland is degraded, as topsoil disappears, as water becomes scarcer and as the energy required to fuel farming becomes more expensive, we will have to consider new ways of feeding ourselves, and the next few billion of us.

By our own estimations, humans are not endangered nor have we yet reached our peak. In many ways, though, we are our own worst enemies. Even while we try to work out ways to feed a sky-rocketing population, we uncover cures for diseases that will see us living longer and using up more of our precious resources. It is only when we consider ourselves as a species – like any other on this planet – that we realize that our time here is finite. Whether we will eventually succumb to a deadly virus, perish in a direct hit from an asteroid or exhaust our earthly resources, we do not know. But, like every living thing on this planet, we are driven by instinct to survive and the fact that we have made it this far gives us hope that we can survive long into the future.

Concept for an urban farm. The block incorporates levels for livestock and crops, saving on land and putting farming at the heart of the city.

← - -

- - →

Despite baby longing, the birth rate in Finland has fallen steeply in the last century and is now below replacement level – the level required to maintain the population. At the opposite end of the scale, the ten countries with the highest birth rates in 2012 were in Africa.

BIG DISCOVERY:
BABY LONGING

"I have had 'baby fever' for a long time... When my 7th birthday approached, my mother asked me what kind of present I wanted. I said that I did not wish for anything but a real baby... Yesterday I went to the drug store to get my contraceptive pills. There was a long queue and I went to the newspaper stand to make time pass, I took a magazine from the pile and it turned out to be Two Plus [a magazine for young parents]. How ironic. I looked at the baby pictures in the magazine and tears came to my eyes. I wanted a baby that badly. (Childless university student, b.1985)"

From "'All that she wants is a(nother) baby'? Longing for children as a fertility incentive of growing importance", Anna Rotkirch, published in the *Journal of Evolutionary Psychology*, 25 October 2007.

In 2007, population researcher Anna Rotkirch at the University of Finland published the results of a survey of 3,000 Finnish women, putting forward her theory for why "baby longing" or "baby fever" sets in when a woman is in her 20s. She explained the overwhelming desire for children that many women feel as being a result of three forces: changes in levels of hormones with age; the desire to care for someone or "need to nurture"; and as part of a strategy of mate selection that tests the commitment of a partner. While men, particularly in Western societies, are often persuaded into fatherhood, most men aged 30–39 say they have found themselves doing some baby longing of their own at some point.

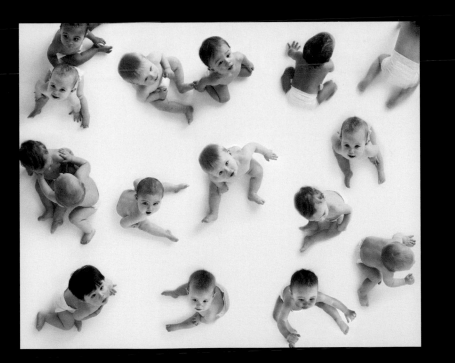

IS TIME TRAVEL POSSIBLE?

As the continents and oceans of Earth drifted beneath him, it is hard to know what cosmonaut Sergei Krikalev must have been feeling. He had been born in 1958 in Leningrad (now St Petersburg), in an unsettled Soviet Union that was still reeling five years after the death of Joseph Stalin. Now his homeland was changing again as he looked on from orbit. Krikalev had departed from Earth and the Soviet Union in May 1991 to spend 10 months on the *Mir* space station. By the time he touched down again in March 1992, the communist state he had left behind had crumbled. He has since been dubbed the "last citizen of the USSR". He is also the greatest time traveller in human history.

Krikalev went on to make subsequent trips aboard the space shuttle and the International Space Station, and by the end of his sixth space flight in 2005 he had racked up a total of 803 days 9 hours and 39 minutes in space – a record. It was during these trips, orbiting the Earth at 28,000 kilometres per hour (17,500 miles per hour), that he time-travelled into the future. By the time his spacefaring days were over, he was 0.02 seconds younger than he would have been if he had stayed on Earth. He had travelled these 0.02 seconds into his own future as a result of time dilation – an effect that was first predicted by Albert Einstein in 1905 as part of his special theory of relativity.

In the decades before Einstein's groundbreaking publication it had been known that light was weird. A famous experiment performed by US physicists Albert Michelson and Edward Morley in the 1880s showed that light always travels at the same speed. This speed, which we know today is exactly 299,792,458 metres per second (186,282.397/05122 miles per second) in a vacuum, remained fixed no matter how the light was observed. This finding ran counter to previous ideas about relative motion. For instance, imagine a sailor on the deck of a ship moving parallel to the shore. He hits a golf ball off the deck out in front of the ship. A bystander on the shore would measure the golf ball to be travelling at the speed of the ship plus the speed of the ball. This is an example of Galilean relativity, formulated by Italian astronomer and mathematician Galileo Galilei in the seventeenth century. But what if the golf ball is replaced with a torch and the sailor shines the torch out ahead of the ship? According to Galilean relativity, the person on the shore would measure the light to be moving at the speed of the ship plus the speed of light. But this isn't actually what happens: the speed of the ship does not add to the speed of the light (which is fixed). In other words, light from a moving source travels at the same speed as light from a stationary source.

The revolution came when Einstein realized that the constant speed of light had implications for time itself. If the observers aren't allowed to disagree on the speed of light, then a way to reconcile things is that they must disagree on the time. The result is that time runs differently for two observers when they are moving relative to each other. A consequence of this is that moving clocks run slow – time is said to "dilate" for a person moving at a constant speed relative to another person. We don't experience

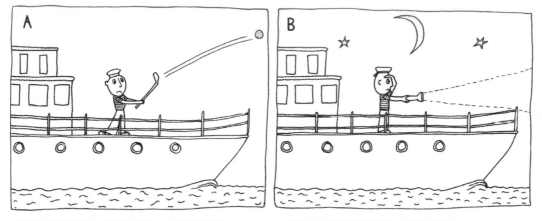

A. SPEED OF GOLF BALL = SPEED OF BALL + SPEED OF SHIP

B. SPEED OF LIGHT FROM TORCH = SPEED OF LIGHT

this time dilation in our everyday lives because the effect is minuscule at the low speeds we encounter day to day. If it weren't, then taking a six-hour flight across the Atlantic would cause us to age noticeably more slowly than those still on the ground. Nonetheless, air travel does dilate time and its very small effects have been accurately measured by atomic clocks flown around the world (*see* BIG DISCOVERY: THE HAFELE-KEATING EXPERIMENT).

The effect of time dilation may be small in normal life but at the speeds travelled by orbiting satellites and space stations – and the astronauts in them – it increases enough to become significant. In fact, without accounting for time dilation, some common aspects of modern life wouldn't be possible. The aircraft that ferry millions of people a year across the planet use the satellites of the Global Positioning System (GPS) to pin down their locations to within 5 metres (about 16 feet). The same technology is used in car satellite navigation systems and map apps on smartphones. On board each of the GPS satellites is an atomic clock accurate to one-billionth of a second. Your position on Earth is calculated by receiving time signals from several of these satellites. If the GPS system didn't take this form of time dilation into account, the satellites would lose time by 7 millionths of a second every day. However, they would also gain almost 46 millionths of a second due to another form of time dilation – gravitational time dilation (*see* INSIDE EXPERT: ARE THERE OTHER WAYS TO TIME TRAVEL TO THE FUTURE?) – so, overall, they would be out by 39 millionths of a second per day. This may not sound much, but it is equivalent to about 11 kilometres (7 miles) of location accuracy on the ground, which is far too large a margin of error when, for example, tracking planes in the busy skies around an airport, so it is fortunate that the GPS system does compensate for time dilation.

So time travel to the future is straightforward – you just need to go really fast. Travel around space at 99.9999 per cent of the speed of light for 10 years and you'll return to Earth at about 9000 CE. Humans have already accelerated tiny subatomic particles to such superfast speeds – the record is 99.9999999999874 per cent of the speed of light, achieved inside particle accelerators like those at CERN near Geneva in Switzerland. Time travel to the future is no longer primarily a scientific problem but the technological one of developing the hardware needed to accelerate much heavier objects, like people, to the extremely high speeds required for Einstein's time dilation to kick in significantly.

In one sense, you are already travelling into the future just by reading this book. Exploiting the effects of time dilation just means you hit the fast-forward button and get there sooner. But what about time travel to the past? For decades, scientists have pondered whether it is possible to reverse the relentless march from past to future and travel in the opposite direction.

THROUGH THE WORMHOLE

One way it might be possible to travel into the past is to use a wormhole – a hypothetical bridge between two parts of space. Imagine a piece of paper. If you want to travel from the top to the bottom you might think you need to cross the entire distance from one end to the other. But if you fold the page in half, then your starting point and your destination now rest on top of each other. You merely need to make a hole between them and travel twice the thickness of the paper to get from one point to the other. Scientists know that mass bends space. According to Einstein's famous equation $E=mc^2$ (in which E is energy, m is mass and c is the speed of light), energy is equivalent to mass, which means energy can also bend space. Physicists have speculated that a huge amount of energy might bend space enough to create a similar shortcut – or wormhole – between two apparently distant places. The openings of the wormhole are known as mouths and the tunnel between them is called the throat. A wormhole, then, is a strange beast with two mouths and one throat.

A real wormhole has never been discovered, although some researchers believe tiny versions might be fleetingly popping into existence all the time. On the minutest of scales, millions of times smaller than an atom, space gets weird, turning into what physicists call a quantum foam. This foam has a sort of loan agreement with nature called the Heisenberg Uncertainty Principle, which stipulates that you can borrow as much energy as you want as long as the amount of that energy multiplied by the time you borrow it for always remains the same. This means you can either borrow a small amount of energy for a long time, or a huge amount of energy for a short time. A sizeable, short-term energy loan might be capable of creating a tiny wormhole.

The trouble is that these wormholes are "virtual" – they pop out of existence almost as soon as they appear due to the need not to default

Russian cosmonaut Sergei Krikalev (b.1958). During his record 803 days in orbit around the Earth, he time-travelled 0.02 seconds into his own future thanks to the effects of time dilation, making him humanity's greatest time traveller.

The satellites of the Global Positioning System (GPS). The atomic clocks on board each satellite have to be corrected for the effects of time dilation; otherwise they would become useless within a day.

BIG DISCOVERY:
THE HAFELE-KEATING EXPERIMENT

When Einstein published his special theory of relativity in 1905 it was just that – a theory. It was his attempt to explain the peculiarities of light. However, for any scientific theory to gain a following it needs evidence to back up its predictions. With time dilation, one of the cornerstone predictions of Einstein's theory, actual evidence of time ticking at different rates was crucial. The first successful demonstration of this came via an experiment carried out in 1938 by US scientists Herbert E. Ives and G. R. Stilwell in which they measured changes in the frequency of light coming from a moving source and used the measurements to determine the amount of time dilation.

However, a famous experiment of 1971 showed that time dilation wasn't restricted to the laboratory. US scientists Joseph Hafele and Richard Keating boarded a flight for a double trip around the globe accompanied by four atomic clocks. These clocks are extremely accurate, using radiation emitted from caesium atoms to count out time – one second has passed after 9,192,631,770 cycles of radiation. Hafele and Keating first flew eastward around the world, travelling faster because they were flying in the same direction as the Earth's rotation on its axis. They then reversed their journey and completed a westward lap of globe, being slowed by the Earth's eastward rotation. When they compared their atomic clocks to those that had been left on the ground, there was a discrepancy in time that matched Einstein's predictions. Today, factoring in these discrepancies is a fundamental part of the Global Positioning System.

↑ Albert Einstein (1879–1955). His special theory of relativity, published in 1905, predicted that time would run slower for people travelling at high speeds, effectively allowing time travel to the future. This time dilation effect is now a very well-tested phenomenon.

on their enormous energy debt. This sort of wormhole cannot be used to travel through space, let alone time, unless we can pay off the debt on the wormhole's behalf. In the future, technology may have developed to a stage where we can isolate one of these virtual wormholes and pump it full of energy, satisfying the terms of the loan agreement and leaving us able to exploit the tunnel to travel backwards in time. There is even a way it can be used to hand yourself a present on Christmas Day and watch yourself unwrap it. To accomplish such a feat requires the combination of both time dilation and wormholes.

On this quest for a unique Christmas, let's say, somewhat ambitiously, that by 2100 we not only have the technology to travel at speeds close to the speed of light, but can also create wormholes. One mouth of the wormhole is on Earth, the other sits 4.2 light-years (about 43 trillion kilometres, or 24 trillion miles) away at the nearest star beyond the Sun, Proxima Centauri. Light, the fastest thing in the universe, would take 4.2 years to travel the conventional route, but by taking the shortcut through the wormhole that distance can be covered by humans in a matter of seconds.

On 15 December 2100, a spacecraft sets off from Earth on a 5 light-year (47 trillion kilometre, or 29 trillion mile) round trip through space, travelling at 99.5 per cent of the speed of light. Attached to the back of the spaceship is the Earth mouth of the wormhole. Because the spaceship is travelling slightly slower than the speed of light, the 5 light-year trip actually takes 5 years and 10 days, and the wormhole mouth is returned to Earth on Christmas Day 2105. You immediately jump in. What time is it when you emerge at Proxima Centauri? Any clock sitting in the throat of the wormhole would have experienced time dilation due to its whizzing through space. At 99.5 per cent of the speed of light, only 6 months would have ticked by inside the wormhole, as opposed to the 5 years the round trip seemed to take from the perspective of those on Earth. This means the date inside the wormhole is 15 June 2101, only 6 months after the wormhole initially left Earth. Because the throat of the wormhole hasn't moved relative to the mouth at Proxima Centauri, it is also the same date there too. You have jumped in on Christmas Day 2105 and emerged on 15 June 2101, becoming the first human being in history to time travel into the past.

You then jump in a spaceship and travel back to Earth the conventional, non-wormhole route at 99.5 per cent of the speed of light. After the 4.2 light-year journey you arrive back to Earth slightly less than 4 years and 3 months later, in early September 2105. Having just arrived back more than three months before you jumped into the wormhole in the first place, you have plenty of time to pick out the perfect Christmas gift to present to yourself just before you take that initial plunge.

This kind of time machine might help to answer one of the nagging questions of time travel: why we aren't inundated with time tourists. If time travel really is invented in 2100, you might expect the time travellers of the day to want to take a holiday in the past. Rather than sitting at home and watching a documentary on the Second World War or the assassination of Kennedy, you could actually travel to 1940s London or 1960s Dallas to watch the events as they occurred. Images of all the major events in history would suddenly have a plethora of future-dwellers lurking in the background. But we don't see this, and the one obvious reason could be that time travel to the past is never invented in the future. Equally, it could be that time travel to the past is invented but it uses the wormhole method above. In this set-up, you can never travel back to a time before the wormhole was created in the first place, so with no wormholes created by humankind to date, we see no time tourists.

It is safe to say that this method of travelling to the past isn't going to happen any time soon. It is immensely speculative and requires huge leaps forward in technology. However, what it shows is that there is currently nothing in the known laws of physics that forbids time travel to the past. The fact that it is implausible, rather than impossible, means that it is an active area of scientific research, with academics considering the theoretical possibilities of time travel, allowing the laws of physics as we

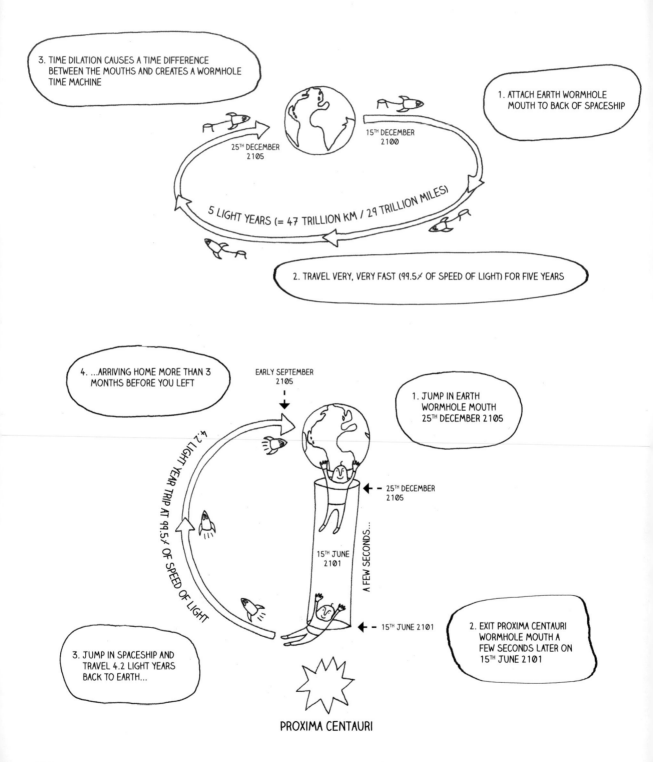

know them to be pushed to their limits. However, some physicists remain extremely uncomfortable with the idea of visiting your own past. What if you were less generous and instead of giving yourself a nice Christmas gift, you shot the other version of you in the chest before they (you) had a chance to jump into the wormhole? Dead, they (you) would no longer be able to jump through the wormhole, reach Proxima Centauri and travel back to be there to shoot themself. So how did the person firing the gun come to be there? This paradox is only one of the many problems with time travel to the past, so those wishing to have dinner with themselves might be in for a long wait. However, fast-forwarding the centuries and seeing the distant future might turn out to be less of a pipe dream. In fact, it might be your only hope of finding the answers to the biggest questions facing scientists today – the universe's great unknowns.

The International Space Station (ISS) in orbit 370 kilometres (229 miles) above the Earth's surface. Over 200 people have visited the ISS since it was launched in 1998. Each of them time-travelled a minuscule amount into their future thanks to time dilation.

INSIDE EXPERT:
ARE THERE OTHER WAYS TO TIME TRAVEL TO THE FUTURE?

"Time dilation doesn't just depend on how you're moving – it also depends on where in the universe you are. A clock placed near to a massive object will run slower than an identical clock positioned further away, where the object's gravity is weaker. This 'gravitational time dilation' is another consequence of Einstein's relativity, which says that being in a gravitational field is equivalent to undergoing accelerated motion. Near an object like the Earth, this means that time passes more slowly the closer you are to the centre of the planet.

For astronauts hoping that their orbital motion will save them precious time relative to the folks on the ground, gravitational time dilation is a bit of a downer. Because they're further from the Earth's centre, the astronauts actually experience less gravitational time dilation than people on the ground, working against the time dilation due to their motion. Luckily, under the influence of the Earth's rather puny gravity, the difference in height between the ground and orbit only amounts to a millisecond difference each year. So the orbital motion will still win out and, overall, astronauts really do age slower than people on the ground. The effect is stronger

the more massive the object and the deeper its gravitational field. In the vicinity of a black hole, the gravitational time dilation is so strong that observers watching from a safe distance would see time appear to slow to a halt for any object at the event horizon, the point of no return beyond which even light can't escape."

Dr Marek Kukula, Public Astronomer,
Royal Observatory Greenwich, UK

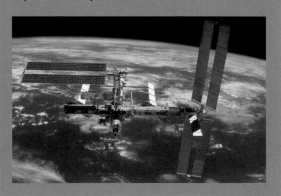

INDEX

Figures in *italics* indicate captions.

Adleman, Leonard 108, 112, 113
age-related diseases 167, 171
ageing 164–71
ageliferin 151
air pollution 83
alcohol 137, 143
Aldrin, Buzz 109
alien hand syndrome 52
alien life 28–37, 152
Allen Telescope Array, California *32*, 35
Alvin 146, 148, 152, 153
Alzheimer's disease 151
amino acids 20, *21*
anaesthesia 52–53
Andromeda galaxy 15, *15*, 16
anti-D mesons 70, 71
anti-neutrinos 72, 73
antibiotics 127–34, 151, *152*
antimatter 67–68, 69, *70*, 72, 73
apes: compared with humans 44
Apple 120
Aristotle 117, 125, 148
Armstrong, Neil 109
artificial intelligence (AI) 56–57, *57*,
    121–22, 123
artificial leaf 91, 92, 94–96, *94*
Aserinsky, Eugene 65, *65*
ASIMO 117–18, *119*
Asimov, Isaac: "Runaround" 122
atomic-scale transistors 110
atomic clocks 184, 186, 189
atoms 9, 13, 17, 67, 75, 78, 79, *96*,
    110, 158
atrophy 169
autonomic nervous system 55
Ava 119–20

baby-longing 176, *180*, 181
bacteria 126–35, 151, 174
bacteriophages 129
barreleye fish 148, 150, *150*, 152
bathyscaphe 145, 146, *146*
Big Bang 10, *12*, 17, 36, 67, 68,
    75–76, 77
binary code 111, 112, *112*
biological computers 112–14, *115*
bioluminescence 148, *148*
Black Death 174
black holes 71, 154–63, 189
Blackburn, Elizabeth 165
Blagosklonny, Mikhail 169
Bohr, Niels 78, 79
Bolden, Charles 177
bosons 159
BRAF mutation 138, 139, 142
brain
    size 42–44, *42*
    weight 42, 50
brain-imaging technology 61
BRCA mutation 138, 139
bristlecone forests 88–89

Caesar cipher 106
calorie restriction 167–68

Cameron, James 145, 150, 151, *151*
cancer 136–43, 167, 169, 170
Cancer Genome Project 138
Cancer Research UK *138*
Capek, Karel: *R.U.R. (Rossum's Universal
    Robots)* 117, 122
car manufacturing 117, *118*
carbon capture and storage (CCS) 84, *85*
carbon dioxide 20, 83–86, *85*, 89, 91
carbon storage 82–89
cardiovascular disease 167, 170
cat in a box experiment 78–79, 80
Cayman Trough 146
cells
    cell division 21, 129, 142, 165
    the first 25, *25*
    growth 169
Census of Marine Life 145, 149
CERN, Geneva 67, 69, *69*, 71, *72*, 185
Chain, Ernst 127
Challenger Deep 145, 146, 150
Chalmers, David 53–54, 55, 57
chatbots 121
chemosynthesis 148
chip technology 109–10, *109*, 111
chlorophyll 92, *93*, *95*, *99*
chloroplasts *93*, *95*
Chomsky, Noam 44
chromosomes 165, *167*
clathrate 85, 86
climate change 83, 84, 86, 88, 149, 179
Clostridium difficile 127, 131
Cocks, Clifford 106
code-makers 106–7
cognitive development 44, 45
coma 53
Coma Cluster 14, *14*
computer speed 109–15
conscious mind 55–57
consciousness 48–57, 60
cookie-cutter shark 148
Copenhagen Interpretation 78–81
coral reefs 86, 151–52
cosmic microwave background 36,
    77, *78*
cosmological constant 9, 12, 13, *13*
CP violation 69–73, *70*, *72*
Crab Nebula *156*
credit cards 101, 106, 107, *107*
Crick, Francis 21
critical density 76
Cronin, James 69, *70*
Curtis, Eugene Paul *112*

D mesons 70, 71
DAF-2 gene (grim reaper gene) 168, 170
DAF-16 gene (sweet sixteen gene)
    168, 170
dark energy 11–13, *11*, *12*, 17
dark matter 13–17, *14*, *15*, *16*, 71
dark silicon 110
Darwin, Charles 19, 39, 46, 139
Dawkins, Richard 46
decoherence 79
deep sea angler fish 146, 148, *148*

Descartes, René 51, *51*
dinosaur extinction 174
disease 83, 180
    age-related 167, 171
DNA 6, 21, 22, *23*, 25–26, *27*, 39–41,
    129, 137, 138, 141, 165, 166–67
    computers 112, 114, *115*
    sequencing 134
Doppler shifts 32
doxorubicin 142
dreaming 52, 58–65
drought 83, 180

$E=mc^2$ equation 185
Earth, origin of 75
Eddington, Arthur *162*, 163
Edwin Smith Papyrus *138*
Ehman, Jerry R. *36*, 37
Ehrlich, Paul 173, 175
Einstein, Albert 7, 9, 12, 13, *13*, 112,
    155, 156, 157, *157*, *158*, 159, *162*,
    163, 183, 185, 186, *186*
electric cars 83
electroencephalography (EEG) 65
electromagnetism 158, 159, 160
electrons 68, 69, 72, 91, 159
ELIZA program 120
Escherichia coli 135
Euclid 101
Euler, Leonhard 101–2, *102*, 103, 107
Europa moon, Jupiter 28, 30, *30*
European Space Agency 29
Everett, Hugh, III 79, 80
evolution 19, 22, 23, *27*, 43, 45, 75, 91,
    129, 147, 169
extrasolar planets (exoplanets) 31, *31*,
    32, 33, 35
extra-terrestrial intelligence (SETI)
    35, 36
extremophiles 152

faecal transplants 131
feeding the world 178–80
fermions 159
fertility 167
fission *96*
Fitch, Val 69, *70*
Fleming, Alexander 127, 129, 130, *130*
floods 83
Florey, Howard 127
flu pandemic (1918) 174, 175, *175*
foraminiferans 86
fossil fuels 83, 84, 89, 91, 94, 99, 178
free-divers 145
Freud, Sigmund 59, 60, 61
fusion *96*, 99, *99*

galaxies 9–12, *11*, 14–17, *15*
Galilean relativity 183
Galileo Galilei 183
General Motors 117
general relativity theory 9, 12, 13,
    155–61, *158*, *161*, 163
genetic code 6, 22, 39
genetic research 21–22, 142

genetically modified (GM) foods
    178–79, *178*
genetics 45, 46
Gilbert, Walter 22, 23
Global Positioning System (GPS) 184,
    185, 186
global warming 83, 88, 89
globalization 130
Goldilocks Zone 30, 31, 35
good bacteria 131
Google 119–20
Government Communications
    Headquarters (GCHQ) 106
graphene 114–15, *114*
graphite *114*
Grätzel, Michael 92
Grätzel cells 92
gravitational lensing 16, *16*
gravitational time dilation 189
gravity 9–12, 16, 17, 29, 75, 155, 156,
    159, *162*, 189
Great Depression 167, *168*
Greenberg, Brooke 165, 166, *170*, 171
greenhouse gases 83, 86, 88

H5N1 outbreak 174–75
Hafele, Joseph 186
Hafele-Keating experiment 186
Hardy, G.H. 105
Heisenberg Uncertainty Principle 185
Helicobacter pylori 143
helium 155
hepatitis B and C 143
heredity 6
Hilbert, David 104, *104*, 105
Hilbert's problems 104
HIV/AIDS 176
Hobson, J. Allan 60, 61
Homer 125
    Iliad 117, 124
Homo erectus 43
Homo habilis 39, *45*
Homo sapiens 45, 75, 173, 175
Honda 117
horizontal gene transfer 129
Hubble, Edwin 9, 75, 76
Hubble Space Telescope 16
Human Genome Project 6, 39, *40*,
    41, 45
human papilloma virus 143, *143*
humanity 38–47
hunter-gatherers 47
hydrogen 91–94, *94*, *96*, 99, 155
hydrothermal vents ("black smokers")
    26–27, 29, 146, 152, 153, *153*
hyperfunction theory 169
hyperplasia 169
hypertrophy 169

ice melting 83, *86*, 87, 89
immortality 170
infinities 157–58, 162
insolation *96*
Intel 109, 110, *110*
International Space Station 183, *189*

International Union for Conservation of Nature and Natural Resources (IUCN) 173
Red List of endangered species 175
internet 10, 104, 106, *106*
iPhone 120
iRobot 119
"Ivy Bridge" computer chip 110

Jobs, Steve *111*
Joint European Torus (JET) *96*, 99, *99*
Jupiter Icy Moons Explorer (JUICE) 29

Kaluza, Theodor *158*, 159–60, 161
Kaluza-Klein theory 159, 160, 161, 162
Kanzi (a bonobo chimp) 39, *40*
kaons 68, 69
Keating, Richard 186
Kenyon, Cynthia 168
Kepler space telescope 32, 35
Kinect (in Xbox games console) 119
Kirkwood, Tom 167
Kiva 117
Klein, Oskar 160, 161
Kleitman, Nathaniel 65, *65*
Krikalev, Sergei 183, *185*

labour, sexual division of 47
lactase gene 46, *47*
language 44, *44*, 46
Large Hadron Collider (LHC) 17, 67, 68, 69, *69*, 73
Leonardo da Vinci 117
LHCb experiment 69–70, 71, *72*
life, origins of 18–27
life expectancy 170
liposomal doxorubicin 142
Lonesome George 165, *166*
luciferins 148, *148*
lysozyme 127

M-theory 159, 162
McCarley, Robert 60, 61
McClintock, Barbara 165
Many Worlds theory 79, 80, 81
Mariner 4 spacecraft 30
Mars *2*
  and alien life 29–30, *30*, 33
  and human habitation 177, *177*
  water on 30, *36*
Mars exploration rovers *92*
Mars One 177
matter 67, 68, 69, *70*, 72
Maury, Louis Ferdinand Alfred 59–60, 64
Maxwell, James Clerk 159
mayfly 165
meat, artificial 178
memory maze 61–62, *62*, 63
metastasis 140
methane 86
Methuselah tree 83, 84, 86, 88, 89
Michelson, Albert 183
Microsoft 119
Milky Way 14–15
Millennium Prize Problems 104
Miller, Stanley 19–20, *20*, 21, 26, 27
minimally conscious state (MCS) 49, 53, 54
minocycline 135
"missing matter" 9
Moeller, Peter 151
Moore, Gordon 109, 110, *110*, 115
Moore's law 109, 110, 112, 115
Morley, Edward 183

MRSA (methicillin-resistant *Staphyllococcus aureus*) 127, *134*
Muller, Hermann J. 165
mutation 129, 137–42, 166–67, 168, 170

nanotechnology 142–43, *143*
NASA 152, 177
nematode worms 167, 168
neurons 42, *42*, 44, 50, 52, 121–22
neutrinos 71–77, 73
neutrons 99
Newton, Isaac 155, 156
nitrogen 20
Nocera, Daniel *94*, 95–96
nucleic acids 22
nutrients 43–44, 169, 178

Oort, Jan 14–15, 16
Oparin, Aleksandr Ivanovich 19, 20
osteoporosis 170
Owen, Adrian 49, 50
oxygen 91, 94, *94*, 96

pandemics 174, *175*
peat bogs 86, 88, *88*
penicillin 127, 130, *130*, 132
persistent vegetative state (PVS) 49, 53
pharmaceutical industry 133
photons 91, 157, 159
photosynthesis 84, 91, 98, 148
  artificial 96
photosynthesizers 147–48
photovoltaics *92*
phrenology *60*
phytoplankton 84–85
Piccard, Auguste *147*
Piccard, Jacques 146, *146*, *147*, 150
Planck length 160
plasmids 129
Poincaré conjecture 104
population *2*, 172–81
positron emission tomography (PET) 65, 67
positrons 67, 68, 69
Post, Mark 178
poverty, extreme *176*
prime numbers 100–107
primordial soup 19–22, *21*, 26
Principe *162*, 163
probiotics 131
progeria 165, 166
proteins 21, 22, *23*, 24–25, 39
protocells 22, 25, *25*
protons 67
Proxima Centauri 186, 187, 189
Pseudomonas aeruginosa 135
public-key (PK) encryption 106, *107*

quantum computers 111–12
quantum mechanics 158–62, *161*
quantum physics 78, 160
quantum suicide experiment 80–81
quantum theory 79
quarks 17, 159
qubits 111, 112

radial velocity technique 32
radio signals 34–37, *36*
radiotherapy 141, 142
red giants 155
redshifts 9, 10, *10*, 14
REM (rapid eye movement)
  dream state 54
  sleep 61, *65*
remotely operated vehicles (ROVs) 146

resistome 129
Riemann, Georg Friedrich Bernhard 102, *102*, 103, 104
Riemann hypothesis *102*, 103–4, *104*, 105, 107
Ring Nebula *156*
Rivers, Ron 106
RNA 21–22, 23–25, *23*, 27
RoboCup *121*
robot butler 116–25
Rodin, Auguste: *The Thinker 54*
Roomba robot vacuum cleaner 124, *124*
RSA algorithm 106, 107, *107*
Rubin, Vera 15–16, *15*

s-bots 125
Sagan, Carl *177*
Schiaparelli, Giovanni 29, *30*
Schrödinger, Erwin 78, 79, 80, *80*
scyllo-inositol 151
sea sponges 151, *152*
seagrass 86, 88, *89*
seas
  acidic 86
  the bottom of the ocean 144–53
  rising 83, 87, 174
self-replicators 21–26
Shamir, Adi 106
sign language 39, *40*
silicon technology 110, *110*, 111, 112, 114, *114*, 115
Siri 120
sirolimus (rapamycin) 170
sirtuin 170
sleep 52, 59, 61, 62, 63, 65, *65*
Smith, Sinclair 14, 15, 16
smoking 137, 143
soil 173
solar eclipse 155–56, *157*, 159, *162*, 163
solar energy *7*, 84, 90–99
solar panels 91–92, *92*, 96, *96*, 97, 176
spacetime 156, 157, *158*, 159
sparticles 70–71
special relativity theory 157, 183, 186
spectrometer 32
spectrum 32, *34*
speech recognition 120
Sperry, Roger W. 51
split-brain patients 51–52
squarks 17
Streptomyces 129, 133, 134
streptomycin 129, 132
string theory, theorists *158*, *161*
subterranean living 177
Sulston, Sir John *40*
Sun
  death of 155
  light-bending 155, 156, *157*, 163
supernovas 155, *156*
  Type 1a 10, *11*
superposition of states 78, 112
superstring theory 159, 161–62
supersymmetry 17, 69–71, *72*, 159, 162
sustainable pod homes 176
Swanton, Charles 139, 140
swarm robotics 124–25, *125*
Swarmanoid 125
symbiosis 135
Symbrion 125–26
Szostak, Jack W. 165

T-cells 142
Tabtiang, Ramon 168
tamoxifen 142
telomerase 166
telomeres 165, 166, *167*, 169
terraforming *2*
theory of everything 158
tidal heating 29
time dilation 183–86, *185*, 189
time travel 182–89
tokamak *96*
toolmaking 39
topsoil 173, 180
TOR pathway 170
tower blocks 176–77
trade 46–47
transits 31, *31*
travelling salesman problem 112, 113
Trieste (a bathyscaphe) 145, 146, *147*
tuberculosis (TB) 127, *128*, 129
Turing, Alan 104, 120
Turing test 120–21
Tyndall, John 6
type-2 diabetes 170

unconscious brain 52–53, *53*
universe
  expanding 9, 10–11, *11*, 12, *12*, 13, *13*, 75–76
  flatness problem 76–77, *76*
  inflation theory 76–77, *76*, 77, *78*
  multiple universes 77, 80, 112
  origin 10
  other universes 74–81
  pocket universes 77–78, *77*
  shape 76, *76*
  stuff in the 66–73
urban farming *180*

vacuum energy 11–12, 13
Venus
  hottest planet in Solar System 30
  Transit of Venus *31*
Virgo Cluster 14
viruses 174–75
volcanoes 26, *26*, 27

Waksman, Selman 132, *132*
Walsh, Don 146, *146*, 150, *151*
Walter, William Grey 56–57
water
  and alien life 29–30, 36
  and livestock farming 178
  shortage 178, 180
  splitting 94, *94*, 95
Watson, James 21, 22
Wayne, Ron *111*
Weizenbaum, Joseph 120
Wellcome Trust Sanger Institute 138
Werner's syndrome 165, 166
white dwarfs 10, *156*
Williams, George C. 167
WIMPS (Weakly Interacting Massive Particles) 17
Winfree, Erik 114
World Health Organization 171
wormholes 185–87, 189
Wow! signal *36*, 37
Wozniak, Steve *111*
Wright, Gerry 135

zeta function 102, *102*, 103, 104
Zwicky, Fritz 14, *14*, 15, 16

CREDITS

The publishers would like to thank the following sources for their kind permission to reproduce the pictures in this book.

Akg-Images: SMB, Gemäldegalerie: 169; **Alamy:** D Hale-Sutton: 80; **American Association for the Advancement of Science**, published Sept 4 1953: 65 (right); **The Bridgeman Art Library:** State Central Artillery Museum, St. Petersburg, Russia: 102; **Corbis:** Ingo Arndt/Minden Pictures: 149 (bottom); /Bettman: 20, 111 (top); /Sonke Johnsen/Visuals Unlimited: 149 (top); /Frans Lanting: 40 (left); /LP7: 47; /NASA/Science Faction: 87; /Ocean: 88 (top); /Radius Images: 98; /Roger Ressmeyer: 110; /Science Picture Co/Science Faction: 45; /Norbert Wu/Science Faction: 150; /Visuals Unlimited: 149 (centre); © **Chris Dwyer:** 115; **EFDA:** 96, 99; **Getty Images:** Aping Vision/STS: 107; /Scott Barbour: 40 (right); /Rodrigo Buendia/AFP: 166; /Digital Vision: 181; /Adam Gault: 106; /Ann Hermes/ The Christian Science Monitor: 124; /E. O. Hoppe/Mansell/Time Life Pictures: 186; /Attila Kisbendek: 123; /Dan McCoy/Rainbow: 42; /Sebastien Micke/Paris Match: 170; /Keystone-France: 130 (right), 132; /Peter Macdiarmid: 138 (bottom); /National Geographic Stock: 148; /Tomohiro Ohsumi/Bloomberg: 119; / Mustafa Ozer/AFP: 121; /Tosh Sasaki: 26; /Roberto Schmidt/AFP: 176; / The Asahi Shimbun: 174; /Johannes Simon: 69; /SSPL: 111 (bottom); / Harald Sund: 54; /Time Life Pictures/NASA: 185 (top); /John Tlumacki/ The Boston Globe: 114; /Universal History Archive: 60, 169; **Courtesy of Martin M Hanczyc:** 22, 24; **Library of Congress, Prints and Records Division:** 128; **Courtesy of Andrei Linde, Stanford University Dept. of Physics:** 77 (top); **Courtesy of NASA:** 30 (left), 189; /AURA/STScI/ESA: 156 (top); /Davide De Martin/ESA/Hubble: 156 (bottom); /ESA, The Hubble Key Project Team and The High-Z Supernova Search Team: 11; / JPL-Caltech: 14, 15, 92 (bottom); /WMAP Science Team: 76, 78; **Naturepl. com:** Brandon Cole: 152; **Figure 2 for Neural Correlates of Consciousness by Mormann & Koch:** reference for graph: 53; **The Ohio State University**

Radio Observatory NAAPO: 37; **Rex Features:** KeystoneUSA-Zuma: 151; /Sipa Press: 85; **Dominick Reuter/dominick@reuterphoto.com:** 94; **Science:** 21; **Science Photo Library:** 159; /Julian Baum: 158; /Brian Bell: 125; /Bildagentur-Online/TH Foto: 97; /CERN: 72; /Clouds Hill Imaging: 130 (left); /Henning Dalhoff: 12 (top); /John Durham: 92 (top); /Equinox Graphics: 161; /ESA/CE/Eurocontrol: 185 (centre); /ESA/DLR/FU Berlin (Neukim) 36; /Cecil H. Fox: 140; /Sam Fried: 88 (bottom); /Mark Galick: 2; /Carla Gottgens/Bloomberg: 179; /Gustoimages: 56 (bottom), 129; /Gary Hinks: 10 (top); /Goddard Space Flight: 31; /Hybrid Medical Animation: 167; /James King-Holmes: 6, 143 (right); /Laguna Design: 23; /Claus Lunau: 157; /David Mack: 27; /Maximilian Stock Ltd: 118; /Hank Morgan: 65 (top & centre); / Dr Karo Lounatmaa: 95, 134; /NASA/ESA/STSCI/W.Colley & E.Turner, Princeton: 16; /N.A.Sharp, NOAO/NSO/Kitt Peak/FTS/AURA/ NSF: 34; /National Museum of Health and Medicine: 175; /NOAA: 146; /David Nunuk: 7; /Royal Astronomical Society: 163; /Emilio Segre Visual Archives/American Institute of Physics: 70 (left & right), 103 (top), 105; / Philippe Plailly: 41 ; /© Punch Limited: 62; /Detlev van Ravenswaay: 30 (right), 77 (bottom); /P Rona/OAR/National Undersea Research Program/ NOAA: 153; /Sciepro: 143 (left); /Dr Seth Shostak: 33; /Take 27 Ltd: 177, 180; /U.S. Navy: 147; /Jim Varney: 141; **The Edwin Smith papyrus:** 138 (top); **Thinkstock.com:** 44, 56 (top), 112; **Topfoto.co.uk:** The Granger Collection: 50 (bottom)

Original illustrations by Claire Evans, www.smallworldanimations.com, © Carlton Books Ltd.

Every effort has been made to acknowledge correctly and contact the source and/or copyright holder of each picture and Carlton Books Limited apologises for any unintentional errors or omissions, which will be corrected in future editions of this book.

ACKNOWLEDGEMENTS

A big thank you to James Wills & co. at Watson, Little, and Gemma Mclagan Ram, Alison Moss and all at Carlton for making this book a reality. Thanks also to our copy editor Martyn Page and illustrator Claire Evans for polishing things up, and all the experts who contributed their thoughts and time.

Hayley would like to thank her co-authors for their enthusiasm, collaboration and inspiration, and Claire for providing an alternative perspective. Her husband Jonny has been The Audience and the support service. Thanks also to Jill, James, Fi & Scampi the dog for sharing their studio space.

Mun Keat would like to thank his co-authors, Hayley and Colin, for letting our curiosity get the better of us and lead to a book; the scientists who gave hours of their time to send me papers, review text, and answer my many questions, particularly Benjamin Thompson, Barry J Gibb, Kat Arney, Alan Winfield, Matt Piper, David Gems, Matthew Hutchings, Ajit Varki, Tim Bayne and Janet Strath. Finally, Mum, Dad, Mun Wei, Pui San for putting up with me always asking "Why?".

Colin would like to dedicate this book to Mum, Dad, Phil and Ruth, whose support is more constant than the stars. Thanks also to Hayley and Mun Keat for making this project such a blast and to Jacob Aron, Sam Gregson and Yuval Grossman for their invaluable help on the chapters. I'm eternally indebted to Nick Babbs, Stuart Clark, Mike Dryland, Rob Edwards, Richard Feynman, Elsie Gray, John Griffiths, Fred Loebinger, Laura Nelhams, Richard Pilgrim, Helen Sharman and my wider family who each in their own way have made this book possible.